Introductory Microbiology

ABOUT THE AUTHORS

Dr. D.V. Pathak, Microbiologist has more than 20 years of research, teaching and extension experience. Presently, he is working at CCSHAU Regional Research Station, Bawal (Haryana). He has about forty research articles in national and international journals, eight review articles, two books and two manuals to his credit. He has been awarded J.V. Bhatt national award two times for best publication in Indian J. of Microbiology by Association of Microbiologists of India and one team award for best research station by CCSHAU, Hisar. He has wide experience of teaching undergraduate classes. His major field of research include biological nitrogen fixation, organic farming and bio-control agents. He has also supervised one ICAR *ad hoc* project on crop residue management.

Mrs. Abha Tikkoo is a Scientist (Soil Science) at CCS HAU, Regional Research Station, Bawal (Haryana), India. She has devoted three decades to agricultural research in the south- west Haryana after doing post graduation. Her work is related to management of brackish waters, significance and judicious of use of fertilizers and integrated nutrient management in sustaining crop yield and soil health. She has made significant contributions towards popularizing the technologies concerning safe use of sodic waters, judicious use of fertilizers, potassium requirement of crops and management of problematic soils. She has contributed in many research projects. She is recipient of IPI-FAI Annual Award, 2010 for outstanding field work on "Promoting Balanced and Integrated Fertilizer Use with emphasis on Potassium". She is a prolific writer and has contributed about 90 publications including research papers in national and international journals, review, booklets, chapters and popular articles.

Dr. Sneh Goyal is working as Prof. and Head in Microbiology Deptt. at CCS Haryana Agricultural University, Hisar. She has taught UG courses to students of Agriculture and Home Science students for several years. Presently she is teaching PG courses. She has published around 70 articles in national and international journals, book chapters and written practical manuals for UG and PG classes. Dr. Goyal received B.Sc. from Kurukshetra University and M. Sc. as well as Ph. D. from CCSHAU, Hisar. She is recipient of Japanese NAKAJIMA-HEIWA post doc fellowship for one year. She is handling number of projects from various agencies. She is expert in soil microbiology and organic matter recycling. She has guided a number of M. Sc. and Ph. D. students.

Introductory Microbiology

D.V. Pathak
Abha Tikoo
Sneh Goyal

2015

Daya Publishing House®

A Division of

Astral International Pvt. Ltd.

New Delhi – 110 002

Published by   : **Daya Publishing House**®
                          A Division of
                          Astral International Pvt. Ltd.
                          – ISO 9001:2008 Certified Company –
                          House No. 96, Gali No. 6,
                          Block-C, 30ft Road, Tomar Colony, Burari
                          New Delhi-110 084
                          E-mail: info@astralint.com
                          Website: www.astralint.com

Sales Office   : 4760-61/23, Ansari Road, Darya Ganj
                          New Delhi-110 002   Ph. 011-23245578, 23244987

Laser Typesetting   : **Rajender Vashist**
                          Delhi - 110 059

Printed at   : **Thomson Press India Limited**

PRINTED IN INDIA

Acknowledgements

As a garland of roses, it is our utmost duty to pay reverence with insightful gratitude to the creator of this World; God. The omniscient and omnipresent with whose inspiration, guidance and fragrance, we were able to write this book. These can not be expressed in words, but felt in heart and are beyond our depictions. We pray that such unending energy and generosity will pour on us in future also.

We duly acknowledge indebtedness to Dr. K.S. Khokhar, Vice Chancellor, CCSHAU, Hisar, Haryana for granting permission for writing this book. The writing of this book would have not been possible without constant consultation and critical perusal of the published books of eminent microbiologists, biochemists, soil scientists and molecular biotechnologists from India and abroad. We also gratefully acknowledge our gratitude to the authors of e-books for monographs, e-sketches and notes; otherwise it would not be possible for this manuscript come into this presentable shape. It is our pleasure to extend sincere thanks to Mrs. Akanksha Tikkoo Singh, daughter of Mrs. Abha Tikkoo for significant and constructive contributions for giving proper shape to chapters and sketches and reviewing the manuscript at all levels.

We are also thankful to Dr. S. S. Yadav, Regional Director, RRS, Bawal, Haryana, for his guidance and cooperation. We extend our sincere thanks to Dr. Manoj Kumar, Research Associate for his assistance in preparation of the manuscript. We would like to take this opportunity to acknowledge the generous co-operation of our colleagues for their advices and helpful comments.

We feel bereft of words in expressing our gratitude to our adorable elders for their everlasting love, blessings, encouragement and motivation. Last but not least, the authors extend sincere thanks to their family members for their love, patience and unaccountable encouragement.

D.V. Pathak

Abha Tikkoo

Sneh Goyal

Preface

The science of microbiology is the study of microorganisms and their activities. It deals with their various forms, structure, reproduction, metabolism, physiology and identification. It also includes their distribution in nature, relations to each others and to living things, their beneficial and detrimental effects and the physical and chemical changes they make in the environment. They are of central importance to major issues of concern to our society; the exploration of life in space, biotechnological tools, genetically engineered microbes involved in production of hormones, enzymes, drugs and other products of pharmaceutical importance. In agriculture, introduction of GM crops resistant to various pests has revolutionized the crop production and crop productivity. It is no exaggeration to claim that there is no biotechnology without microbes. All the geochemical transformations are brought about by microbes and hence, we can say with certainty that there can be no life possible without microbes. There is another side of the coin as all the dreadful diseases are also caused by microbes.

This book entitled "Introductory Microbiology" provides innovative information on the basic and applied aspects of microbiology to the students of undergraduate classes as this book has been written keeping in view for the students having their first academic experience with the world of microbes. It illustrates all important points diagrammatically, and where appropriate the illustrations attempt to give some feeling for the scale and relationships of the elements involved. Most of the reference books available in Indian market are of foreign authors. Language in these text books is difficult to grasp for students of undergraduate level. We hope that the material provided in the book is very precise, logical, to the point, easy to grasp and provide first hand information on all the topics of Microbiology.

We also hope that this book will make this rapidly advancing subject more accessible and readily allow readers to proceed to more advanced work like nanotechnology, bio-informatics, bio-medical engineering etc. Main emphasis in the book has been given to bacterial classification, role of microbes in biogeochemical cycles with special reference to biological nitrogen fixation, fermentation technology and food microbiology. There is always scope of improvement; hence positive criticism from readers is most welcome.

D.V. Pathak

Abha Tikkoo

Sneh Goyal

Contents

Chapter-1

History of Microbiology

Microbiology (from *Greek mîkros*, "small"; *bios*, "life"; and *-λογί α , -logia*) is the study of *microscopic organisms*, either *unicellular* (single cell), *multicellular* (cell colony), or *acellular* (lacking cells). The organisms called microorganisms are too small to be perceived clearly by the unaided human eye. The organisms with a diameter of 1.0 mm or less are microorganisms and fall into the broad domain of microbiology. Microorganisms have a wide taxonomic distribution; they include some metazoan animals, protozoa, many algae and fungi, bacteria and viruses.

The existence of microorganisms was hypothesized for many centuries before their actual discovery. The existence of unseen microbiological life was postulated by *Jainism* which is based on *Mahavira*'s teachings as early as 6th century BCE Paul Dundas notes that Mahavira asserted existence of unseen microbiological creatures living in earth, water, air and fire. *Jain scriptures* also describe *nigodas* which are sub-microscopic creatures living in large clusters and having a very short life and are said to pervade each and every part of the universe, even in tissues of plants and flesh of animals. The *Roman Marcus Terentius Varro* made references to microbes when he warned against locating a homestead in the vicinity of swamps "because there are bred certain minute creatures which cannot be seen by the eyes, which float in the air and enter the body through the mouth and nose and there by cause serious diseases". In 1546, *Girolamo Fracastoro* proposed that *epidemic diseases* were caused by transferable seedlike entities that could transmit infection by direct or indirect contact, or vehicle transmission. However, early claims about the existence of microorganisms were speculative, and not based on microscopic observation. Actual observation and discovery of microbes had to await the invention of the *microscope* in the 17th century.

Theory of Spontaneous Generation

People believed that microorganisms arose spontaneously from non living materials. They were produced *de nova;* this belief of living things that they were produced from non living things was known as theory of spontaneous generation or abiogenesis, which existed for a long time. This theory received a big boost when Needham supported it due to his faulty experiments. He heated meat and left uncovered due to which maggots developed in the putrefying meat.

Controversy over spontaneous generation

Three hundred years ago, there were two distinguished groups who had diverse views over the origin of life. While the one group believed in the spontaneous generation of living beings from non-living matter, the other group advocated the view that life begins from life only. As the science began to grow, proofs accumulated which disproved the theory of spontaneous generation. In fact, the belief of spontaneous generation was the result of inadequate observations and faulty experiments. However, in scientific world it was necessary to disprove the theory of spontaneous generation and to draw any relations between microorganisms and disease or fermentation.

Contributions of different Scientists in the field of Microbiology are mentioned below in brief

1. Antonie Van Leeuwenhoek (1632-1723)

The discoverer of microorganisms or microbial world was Antonie Van Leeuwenhoek; a Dutch merchant by profession He called them animalcules.

He was a Dutch merchant and his hobbies were to grind glasses and make lenses. He constructed simple microscope capable of giving magnification up to 50-400 times. He described some of the unicellular microorganisms like protozoa, algae, yeast and bacteria so accurately that they are not different from what we know about them today. He is also known as Father of Microbiology.

2. Francesco Redi (1626-1697)

He was the first scientist to challenge the theory of spontaneous generation. Earlier the developments of maggots in the meat undergoing putrefaction were believed to be spontaneous. Redi showed that the development of maggots was not spontaneous from putrefied meat. Maggots were the larval stage of flies which laid eggs when meat was kept open in the air. He showed that if it was covered with a paper of fine wire gauze, there was no development of maggots; because flies could not have the access to meat.

3. Louis Joblot (1711)

The development of microbes in hay infusion was considered to be spontaneous. Joblot found that hay infusion prepared by boiling and tightly stopper immediately; remained free of microorganisms. However when these stopper preparations were later opened and exposed to the air, the animalcules soon appeared. This was the first practical proof which ruled out the spontaneous origin of microorganisms.

4. John Needham (1731-1781)

He was the supporter of the theory of spontaneous generation. He repeated Jablot's experiments. In fact he followed inadequate method for heating and preparation of hay infusion due to which microbes developed. Thus he declared that microbes arise spontaneously and heating of infusion or covering of vessel cannot stop their development.

5. Abbe Lazzaro Spallanzani (1722-1799)

Sixteen years later, Spallanzani reinvented the findings of Needham and found the faults in Needham's experiments. He repeated the experiments of Needham in systematic manner and found that heating prevented the growth of microorganisms. He also showed that there was a direct correlation between microorganisms and spoilage of food material. This was proved by the following experiment. He heated meat in two glass containers. He sealed one container and left the other unsealed. The sealed container was not spoiled while the other was spoiled. People did not agree to this experiment saying that air is required for the growth of microorganisms. By sealing, air was not provided.

6. Theodor Schwann (1810-1882)

Schwann proved that air contains microorganisms. He passed air through glass tube with cotton inside. He removed cotton, dissolved in alcohol mixture and observed microorganisms under microscope. He disproved the theory of spontaneous generation with the help of this experiment. He prepared swan neck flask. He boiled meat infusion in these flask and allowed air to enter through bent tubes. Meat infusion was not spoiled because microorganisms contained in the air were trapped in bent tubes and air reaching infusion was sterile. Infusion was spoiled when the liquid of flask was allowed into bent neck and poured back into the flask. When the neck was broken, air was allowed to enter directly and the infusion was spoiled. These experiments clearly proved that air contains microorganisms. He is also known for invention of cotton plug technique to avoid spoilage of food material.

7. John Tyndall (1820-1893)

He proved that dust contain microorganisms. He filled flasks with meat infusion, boiled and placed the flasks, without plug in dust free atmosphere. Infusion was not spoiled for a long time. He concluded that

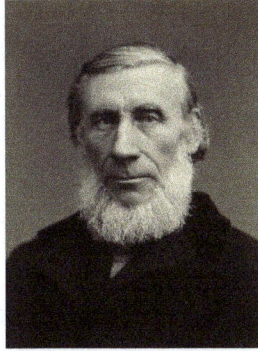

growth did not occur because air did not contain microorganisms. Tyndall observed that some infusions could not be kept sterile by boiling only once. They should be boiled 2-3 times for safe storage. He proved that by cooling intermittent, they remained sterile. Some microbes have two stages in their life cycle: thermo resistant and thermo sensitive stages. Today we term the sensitive stage as vegetative stage and thermo resistant stage as the spores. Due to intermittent cooling, the spores germinated to form the vegetative stage. This vegetative stage was killed at higher temperature. This process of discontinuous heating is known as tyndallization. The process of periodic boiling/ steaming with 2-3 cooling periods (about 1-2 hrs) is called tyndallization.

8. Louis Pasteur (1822-1895)

Louis Pasteur convinced the scientific world that fermentation is the result of microbial activity. Fermentation is a process where microorganisms grow in the absence of air and produce alcohol or acid and carbon dioxide. During fermentation microbes can exist in the absence of air (oxygen). Presently, they are known as anaerobes. These anaerobes were killed when exposed to air e.g. butyric acid fermentation. He evolved a method or technique to preserve food material without loss of quality. It is known as pasteurization, after his name. It is extensively used for preservation of liquid food like juice and milk. It is of two types:

(a) **HTST:** High temperature Short time -161°F or 71.6°C for 15 seconds

(b) **LTLT:** Low temperatue Long time - 145°F or 62.8°C for 30 minutes

In pasteurization only vegetative forms of bacteria are killed. He also evolved methods to immunize people against diseases.

9. Robert Koch (1843-1910)

He gave evidences to show that microbes cause diseases in animals, known as Germ theory of diseases. He found rod shaped bacteria in the blood of rats suffering from disease anthrax. He formulated certain postulates on the basis of his observations which are known as Koch's postulates or germ theory of disease microbes. The postulates are as follows:

- Microorganisms must be present in every case of disease.
- Microorganisms must be isolated from diseased animals and be grown in pure culture.
- The specific disease must be reproduced when a pure culture of microorganisms injected to healthy susceptible animal.
- The microorganisms must be recovered and re-isolated once again from the experimentally injected animal.

In this way, he discovered cause of disease as microbes are responsible for causing diseases like typhoid, tuberculosis and tetanus etc. In addition, he developed technique or methods for isolation, cultivation and maintenance of microbes. On the basis of his work, he got Noble prize in 1905 in physiology and medicine.

The establishment of role of microorganisms in carbon, nitrogen and sulfur cycles was the result of work by S. Winogradsky (Russian) and M. W. Beijerinck (Dutch –Holland). Winogradsky discovered that bacteria use inorganic compounds as energy source e.g. Nitrifying bacteria- *Nitrosomonas*, sulfur oxidizing bacteria- *Thiobacillus*. Beijerinck isolated bacteria fixing atmospheric nitrogen e.g. *Azotobacter*. They also developed a technique for isolating soil microbes known as enrichment technique.

Discovery of antibiotics

- Alexander Fleming discovered penicillin in 1929.
- Streptomycin was discovered by S. A. Waksman in 1945 from actinomycetes named as streptomyces.

Discovery of viruses

- D. Iwanowosky (1892) was the first to observe that extract of tobacco plant with mosaic disease retained infectivity after passing through bacterial proof filters. He discovered that infective agents were smaller than bacteria.
- Loffler and Doetch observed similarly with foot and mouth virus of cattle.
- W. Stanley (1935) showed that TMV (Tobacco Mosaic Virus) was composed of proteins and nucleic acid.
- F. W. Twort and F. d'Herelle (1915) independently discovered that bacteria are susceptible to infectious agents which pass through bacterial filters. Now they are known as bacteriophage.

Microbial discoveries in brief

- The discoverer of microbial world was a Dutch merchant, Anton van Leeuwenhoek (1632-1723). He is known as the Father of Microbiology.
- He developed simple microscope. He discovered the microbial world, the world of "animalcules" or little animals, as he and his contemporaries called them.
- Robert Hooke discovered compound Microscope.
- Francesco Redi (1665) gave the first blow to the doctrine of spontaneous generation, or abiogenesis. He showed that maggots of flies could not develop in the putrefying meat, if the meat was protected with fine gauze as flies were unable to deposit their eggs on it.

- L. Spallanzani concluded from his experiments that an infusion can be rendered barren, if it is sealed hermetically and boiled.

- Francois Appert found that one can preserve the food by enclosing them in air tight containers and heating them. It is known as appertization.

- Louis Pasteur, a French Scientist gave swan neck experiment. In such a flask putrefying materials could be heated to boiling after the flask was cooled, air could renter but the bends in the neck prevented particulate matter, bacteria, or other microorganisms from getting into the main body of the flask.

- John Tyndall in England and F. Cohn TW Germany discovered two forms of microorganisms. The heat sensitive forms were called the vegetative stage, while the heat resistant forms, called as endospares were formed during unfavorable conditions, particularly may infusions.

- John Tyndall gave a technique for billing heat resistant structures called tyndallization. It is the process of discontinuous heating in which the material first heated and than cooled down to convert endosperes into vegetative cells and than again boiling to kill germinated vegetative cells.

- Louis Pasteur also developed vaccines for diseases like anthrax, fowl cholera and rabies.

- Robert Koch used gelatin as solidifying agent for various nutrient fluids. But it had several drawbacks including that it does not remain solid at body temp. (37°C). Further, as gelatin is the protein, hence some bacteria utilize it as nitrogen source.

- The use of agar, a polysaccharide, as the solidifying agent for bacteriological culture was made by Walter Hesse, an associate of Koch.

- In 1887, Richard Petri modified Koch's flat plate technique by the development of double sided dishes after his name.

- In 1798, Edward Jenner developed first vaccine against small pox.

- Robert Lister (1867) developed antiseptic principles in surgery.

- Christian Gram (1884) developed Gram Staining method to distinguish them on the basis of staining.

- Alexander Flemming (1929) discovered first antibiotics.

Chapter-2

Introduction and
Scope of Microbiology

The science of microbiology is the study of microorganisms and their activities. It is concerned with their form, structure, reproduction, physiology, metabolism and identification. All living cells are basically similar. They are composed of protoplasm (Greek meaning "first formed substance"), a colloidal organic complex consisting largely of protein, lipids, and nucleic acids; all are circumscribed by limiting membranes or cell walls; and all contain nuclei or an equivalent nuclear substance.

Why microorganisms are used as tools of biological research?

1. They can be grown in test tubes/flasks or on Petri plates, thus requiring less space and maintenance than large plants and animals.

2. They grow rapidly and reproduce at an unusually high rate; some bacteria undergo almost 100 generations in 24 hrs.

3. The metabolic processes of microbes follow patterns that occur among higher plants and animals.

The science of microbiology is the study of microorganisms, their form, structure, reproduction, physiology, metabolism, identification and their relationship to each other and to other living beings. Microorganisms are too small to be seen by naked eye. Roughly speaking microorganisms are organism with a diameter of 1mm or less; which cannot be seen by naked eyes.

Related areas of Microbiology

The subject of Microbiology can be divided into five sub groups namely:

1. Protozoology
2. Algalogy
3. Mycology
4. Bacteriology
5. Virology

1. Protozoology: It deals with the study of protozoa. Protozoa are unicellular organisms, devoid of cell wall, contain cell membrane, and cause diseases in man and animals e.g. *Amoeba, Paramecium, Plasmodium* (malarial agent), *Trypanosoma* causes sleeping sickness and *Trichonympha*, present in the intestine of termites, converts cellulose into glucose.

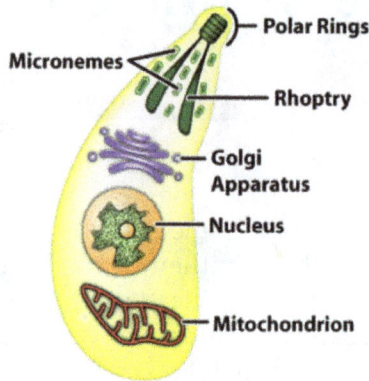

Fig. 2.1. Structure of Protzoan

2. Algalogy: It is the study of algae. Algae may be unicellular or multi-cellular having filamentous structure. It contains chlorophyll and other pigments. It is photosynthetic in nature. Cell wall is made up of cellulose and pectin's. It multiplies by sexual and asexual reproduction. Blue green algae contain phcocyanins, a blue pigment and fix atmospheric nitrogen e.g. *Anabaena* and *Oscillatoria*, Agar agar is extracted from *Gelidium* (Red algae), *Chlorella* is used as food or single cell protein in Japan, *Cephaleuros* causes red rust disease in coffee and tea.

Fig. 2.2. A view of algae

3. Mycology

It deals with the study of fungi. Fungal hyphae are unicellular to multi-cellular, filamentous and non-photosynthetic mycelia. Mode of living is either parasitic or saprophytic. They reproduce sexually as well as asexually. Some causes disease in plants and others are useful ones. Various examples are as follows:

1. *Alternaria* (Blight and leaf spots), *Puccinia* (Rust in wheat)
2. *Aspergillus, Penicillium and Rhizopus*: They spoil food and food products like bread and fruits and fruit juice.
3. *Penicillium*: It is used in production of penicillin antibiotic.
4. *Saccharomyces*: It is used in alcohol, beer, whisky and wine production.
5. *Aspergillus:* It is used in production of citric acid.
6. *Fusarium:* It synthesizes plant growth promoting substance like gibberlins.
7. *Aspergillus* **and** *Trichoderma*: They take part in composting.

Fig. 2.3. Fruiting bodies of Fungi

4. Bacteriology: Unicellular, microscopic and generally non-photosynthetic bacteria are so small that a special unit of measurement (Micron- 1/1000) is used while referring to their size. They constitute one of the largest groups. They are cosmopolitan in nature. Many are very useful while a few are pathogenic to plants and animals. Various examples are as follows.

1. *Rhizobium, Azotobacter and Azospirillum* fixes atmospheric nitrogen.
2. *Thiobacillus* convert soil minerals (P, S, Mn and Zn) from unavailable form to available form to plant and thus increase soil fertility.

3. *Pseudomonas* and *Bacillus* convert organic form of nutrients to inorganic form and thus make them available to plants.

4. *Streptococcus* and *Lactobacillus* helps in producing dairy products like butter, curd and cheese etc.

5. Certain compounds like vitamins (*e.g. Propionibacterium*), amino acids (*e.g. Corynebacterium*); enzymes like protease and amylase are also produced from cheap raw materials.

6. Bacteria are involved in biogas production from cattle dung and other agricultural wastes *e.g. Methanobacterium* and *Methanococcus*.

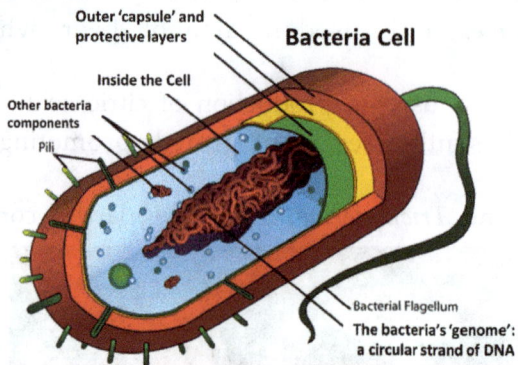

Fig. 2.4. Structure of bacterial cell

5. Virology: It deals with the study of viruses. Viruses are ultramicroscopic forms that can be seen only by electron microscope. They are obligate parasites on plants, animals and microorganisms like bacteria and fungi. Viruses of plant are known as plant viruses e.g. Tobacco Mosaic Virus (TMV), mosaic virus in pulses and vegetables. Viruses of animals are known as animal viruses e.g. pox virus, rabies virus etc. Viruses of bacteria and fungi are known as bacteriophages and mycophages, respectively.

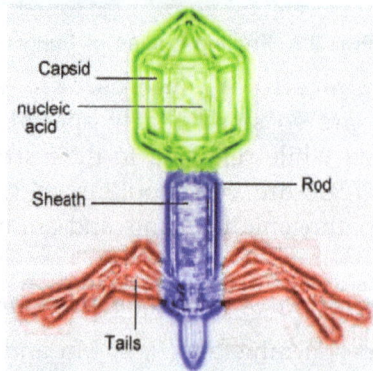

Fig. 2.5. Structure of a Virion

IMPORTANCE OF MICROBIOLOGY

Microorganisms play an important role in our daily life. They are said to be responsible for maintenance of life on this planet. Some contributions of microorganisms are summarized below:

- **Recycling of carbon and other elements:** Green plants utilize energy of sunlight and produce organic matter from CO_2 and thus are called primary producers. Animals utilize the plant biomass and thus are called as consumers. Air contains approximately 0.03 per cent CO_2. The rate of photosynthesis is such that whole of CO_2 of atmosphere will exhaust in about 20 years if CO_2 is not returned back to atmosphere. The microorganisms (bacteria and fungi) carry out this important function of maintenance of life on earth. These microbes function as degrading agents of dead organic matter and return the carbon of organic matter as CO_2 into the atmosphere. Similarly, other elements are also returned back to the atmosphere by degradation.

- **Biological nitrogen fixation:** Nitrogen is an important cell constituent. The yield of crops is dependent on the supply of nitrogen. Nitrogen is provided to the plants in inorganic form. The nitrogenous fertilizers are prepared in big fertilizer plant where atmospheric N_2 is chemically fixed to form ammonia. Chemical fixation of nitrogen is approx 40 x 10^6 tonnes per annum. Nitrogen to the plants is also supplied through biological fixation where certain bacteria fixes atmospheric nitrogen either in association with plant or independently. The biological fixation of nitrogen (170 x 10^6 tones) is far more than the chemical fixation. Various nitrogen fixing bacteria are available as bio-fertilizers for inoculating various food and oil seed crops e.g. *Rhizobium, Azotobacter* and *Azospirillum*.

- **Mycorrhiza:** The roots of many plants are closely associated with fungi called **Mycorrhiza**. The fungi penetrate into cortical cells, form vesicles and arbuscles. The fungi help in absorption of minerals (phosphate and nitrogen) from soil for the plant and help in absorption of water by increasing surface area of roots. On the other hand, plant provides carbohydrates and other nutrients for growth of fungus.

- **Degradation of cellulose in rumen:** Ruminants feed on hay, straw and grasses, of which more than 50 per cent is cellulose. The cellulose component of their food would be useless if it is not degraded. In the course of evolution, the ruminants have developed a symbiotic relationship with microorganisms which are able to degrade cellulose. Protozoa and bacteria are predominant microorganisms of rumen. Bacterial population is in the range of 10^9-10^{10} cells per ml of rumen fluid. The examples of cellulose degrading bacteria are *Ruminococcus*, *Clostridium* and *Cellulomonas*.

- **Silage production:** Silage is the method of preservation of cattle feed with its characteristic flavors, taste and nutritive value. In this, leaves of green plants (Maize, grass and Lucerne) are compacted and molasses is added. Lactic acid bacteria develop and produce lactic acid which helps in conservation of cattle feed e.g. *Lactobacillus* and *Leuconostoc*.

- **Production of Biogas:** Biogas fermenters are used to ferment animal excreta along with cellulose containing crop residues. In this process, nitrogen of the excreta is preserved in the rotting sediment and methane gas formed can be used as fuel for domestic use e.g. *Methanobacterium, Methanococcus*.

- **Composting:** Role of microorganisms in organic matter decomposition and composting is well known. The microbes transform wastes into nutrient rich manures which not only enrich soil but also improve its physico-chemical and biological properties. Organisms like *Bacillus, Pseudomonas, Aspergillus, Penicillum* and *Thermoactinomyces* play important role in composting.

- **Industrial uses of microorganisms:** Some of microorganisms are being used for various industrial purposes *viz;* beer and wine production (*Saccharomyces cerevisiae*), bread making (*Saccharomyces cerevisiae*), milk and dairy products (*Lactobacillus acidophilus* and *Streptococcus lactis*) and formation of vinegar by acetic acid bacteria (*Acetobacter aceti*) since ages. For last six to seven decades, microorganisms have also been used in industrial processes like production of antibiotics, vitamins, harmones and other microbial products.

- **Production of other products:** Penicillin, a bio-product of *Penicillium chrysogenum* was discovered by Alexander Fleming in 1929. This was the beginning of the chemotherapy and pharmaceutical industry. Since then, many antibiotics produced by microorganisms are known as erythromycin, streptomycin, neomycin, tetracycline, polymyxin , bacitracin etc. Above all, carotenoids and steroids, enzymes like amylase, protease and pectinase; and amino acids like glutamic acid can also obtained from microorganisms. Microorganisms are also known to control the pests e.g. *Bacillus thuringenesis*.

General properties of micro-organisms

Microorganisms include diverse group like algae, fungi, protozoa, bacteria and viruses. All of them are microscopic. But dimensions are not the only motive for placing them in a special group separate from plants and animal kingdom. They differ in morphology, activity, diversity, flexibility of metabolism and ecological distribution. Some of the peculiar properties are mentioned below:

1. Surface to volume ratio: The ratio of surface to volume is very large. The volume of medium sized bacterium is 1.0 μm^3. The high surface to volume ratio is the reason for high metabolic rates of microorganisms.

2. Metabolic flexibility: In microorganisms, there is metabolic flexibility. Enzymes are produced by microorganisms whenever appropriate substrate is present.

3. Distribution: These are ubiquitous, present in soils, atmosphere, hot springs as well as in arctic conditions.

Chapter-3

Cellular Organization in Bacteria and Eukaryotes

The common features of biological systems are as follows:

- All organisms share a common chemical composition.
- All organisms perform certain common chemical activities.
- All organisms share common physical structure.

Chemical composition: Three types of complex organic macromolecules present are protein, deoxyribonucleic acid (DNA) and ribonucleic acid (RNA). Proteins are biological catalysts or enzymes. These are involved in different types of biological reactions. DNA is the genetic material and RNA is responsible for carrying message for various microbial activities via protein synthesis.

Metabolism: It involves the processes required for growth and to generate energy which are necessary for their activities.

Physical structure: All organisms are organized into microscopic subunits known as cells. As a result of their cellular organization, growth results in cell division with an increase in total number of cells.

Types of cellular organization of microorganisms

There are two types of cellular organization among microorganisms,

- **Unicellular organization:** They consist of small microscopic single cells. Most common examples are bacteria and protozoa.

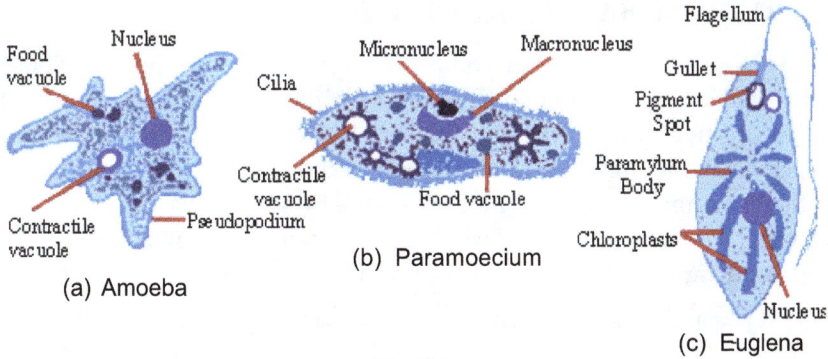

Fig. 3.1

- **Multi-cellular organization**: They consist of many cells attached to one another in a characteristic fashion. Multi-cellular organization arises initially from single cells which is quite common in fungi and algae.

Internal organization of cell in microorganisms

All cells can be grouped into two types: Eukaryotic cell and Prokaryotic cell.

1. Eukaryotic cell: Eu means true and Karyotic means containing nuclear material. Eukaryotic cells means cell containing true nuclear material. Cells are more complex in nature and are commonly found in protozoa, algae, fungi, plants and animals.

2. Prokaryotic cell: Pro means like and karyotic means containing nuclear material. Prokaryotic cell means cell containing nuclear like material. In prokaryotic cell, nucleoplasm is not enclosed by nuclear membrane. Prokaryotic cells are commonly found in bacteria and blue green bacteria.

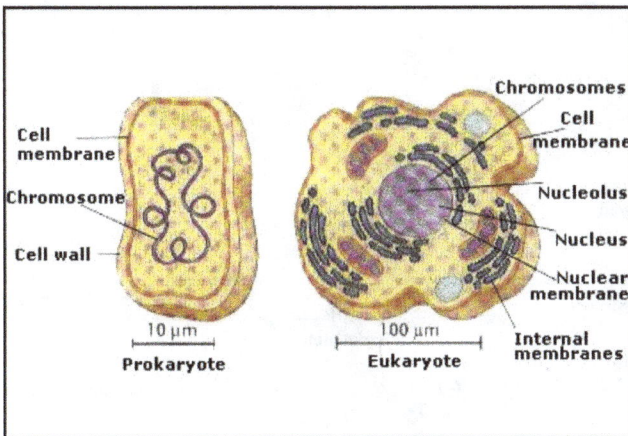

Fig. 3.2. Prokaryotic and Eukaryotic cells

STRUCTURE OF BACTERIAL CELL

The cytoplasm of bacteria contains polysomes - a range of ribosomes actively translating messenger RNA into proteins. Some bacteria also have inclusion bodies within the cytoplasm. These are often energy storage resources. Some inclusion bodies are referred to as metachromatic granules since they change the colour of dyes used to stain cells. Inclusion bodies found within *Corynebacterium diphtheriae*, the causal organism of diphtheria, are an important example of meta-chromatic granules.

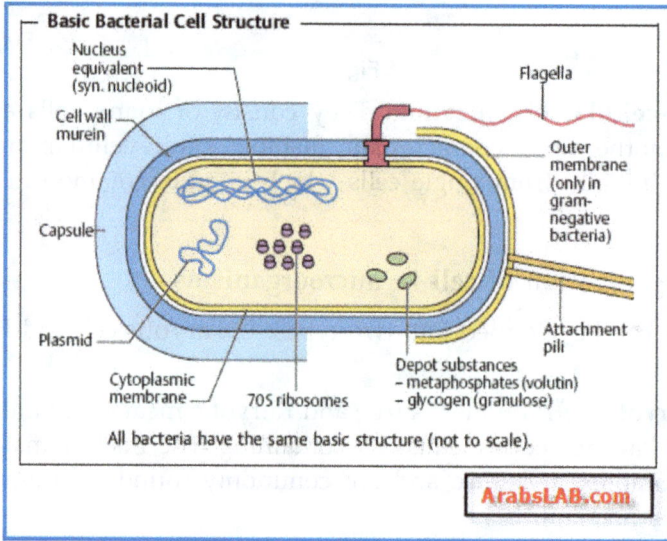

Fig. 3.3. General view of a bacterial cell

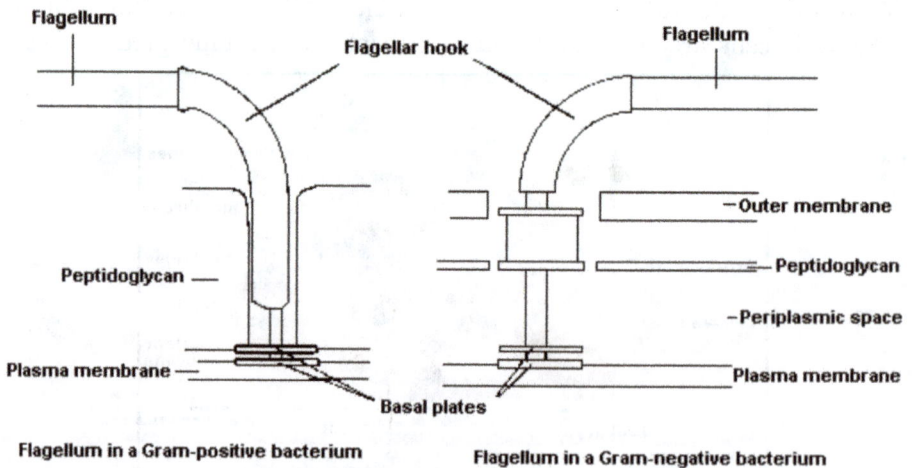

Fig. 3.4. Insertion of flagella into bacterial cells of different types

a Gram positive

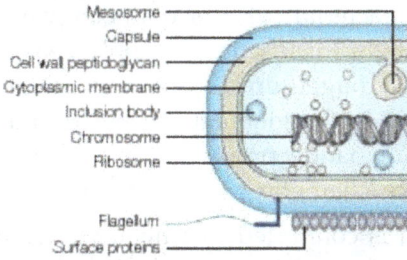

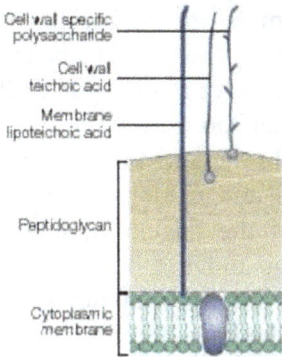

Mesosome
Capsule
Cell wall peptidoglycan
Cytoplasmic membrane
Inclusion body
Chromosome
Ribosome

Flagellum
Surface proteins

b Gram negative

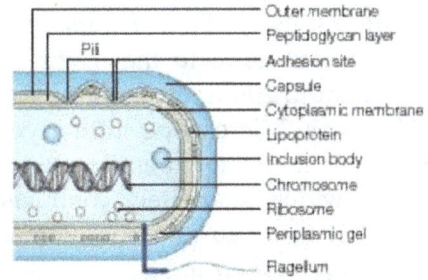

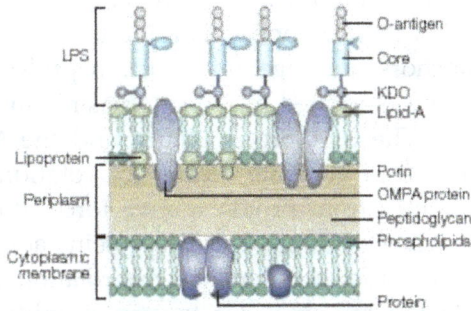

Outer membrane
Peptidoglycan layer
Pili
Adhesion site
Capsule
Cytoplasmic membrane
Lipoprotein
Inclusion body
Chromosome
Ribosome
Periplasmic gel
Flagellum

Cell wall specific polysaccharide
Cell wall teichoic acid
Membrane lipoteichoic acid

Peptidoglycan

Cytoplasmic membrane

LPS

O-antigen
Core
KDO
Lipid-A

Lipoprotein
Periplasm

Cytoplasmic membrane

Porin
OMPA protein
Peptidoglycan
Phospholipids

Protein

Fig. 3.5. Gram positive and Gram negative Bacterial Cell

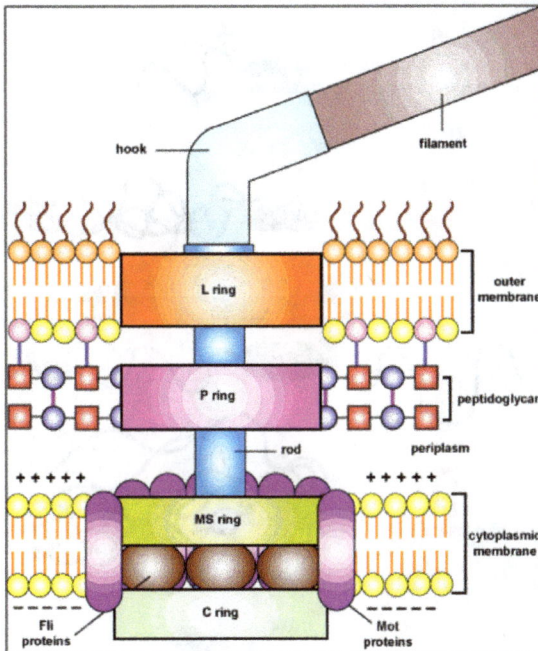

hook
filament

L ring

outer membrane

P ring

peptidoglycan

rod
periplasm

+ + + + +   + + + + +

MS ring

cytoplasmic membrane

C ring

Fli proteins
Mot proteins

Fig. 3.6. Flagellum of Gram – ve Bacteria

Flagellum is composed of three parts - basal structure + hook like structure + long filament outside the cell. The length of the flagellum is several times more to the length of the bacterial cell. Its diameter varies from 10nm to 20nm. It cannot be seen by light microscope, can only be seen by Electron Microscope after dyeing the flagellum. The protein which constitutes it is known as flagellin. Protein is the globular protein forming a chain which alternately contracts and expands producing a wave like motion. It also causes tactic moments like chemotaxis or phototaxis.

The filament of the bacterial flagellum is connected to a hook which, in turn, is attached to a rod. The basal body of the flagellum consists of a rod and a series of rings that anchor the flagellum to the cell wall and the cytoplasmic membrane. In gram-negative bacteria, the L ring anchors the flagellum to the lipo-polysaccharide layer of the outer membrane while the P ring anchors the flagellum to the peptideglycan portion of the cell wall. The MS ring is located in the cytoplasmic membrane and the C ring in the cytoplasm. The Mot proteins surround the MS and C rings of the motor and function to generate torque for rotation of the flagellum. Energy for rotation comes from proton motive force. Proton moving through the Mot proteins drives rotation. The Fli proteins act as the motor switch to trigger either clockwise or counterclockwise rotation of the flagellum and to possibly disengage the rod in order to stop motility.

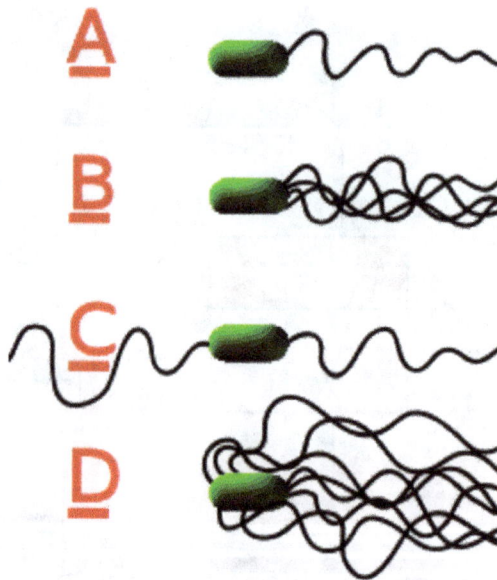

Fig. 3.7. Flagellum

Different types of flagella arrangements:

A. Monotrichous: Single flagellum at one end of cell structure.

B. Lophotrichous: Group of flagella at one end o cell structure.

C. Amphitrichous: Flagella at both end of the cell surface.

D. Peritrichous: Flagella all over the cell surface.

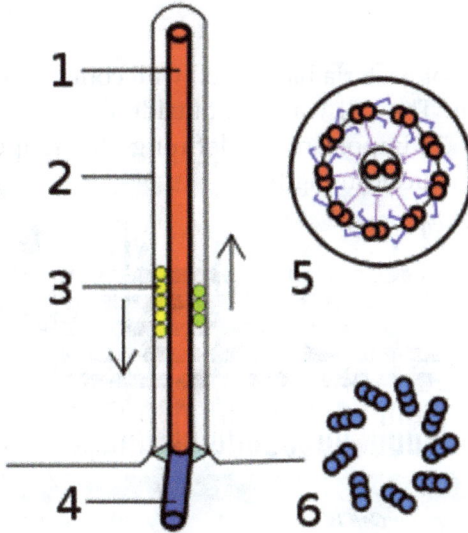

Fig. 3.8. Structure of Eukaryotic Flagellum

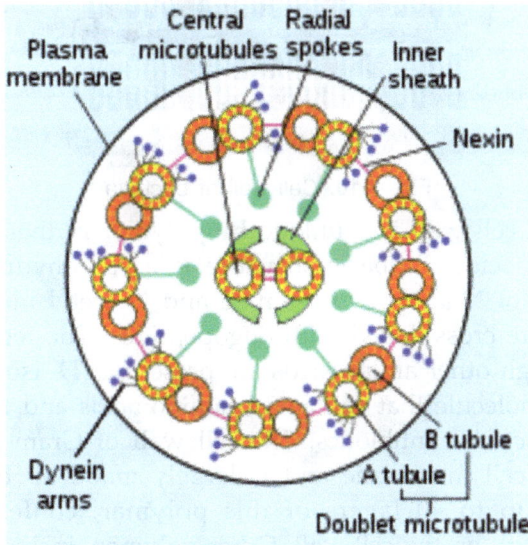

Fig. 3.9. Cross section of an axoneme

Pili (fimbriae)

The appendages, singularly called *pilus* or *fimbria*, are smaller, shorter and more numerous than flagella. They perform several functions, like **F-pilus** or **sex** *pilus* for transfer of genetic material from donor bacterial cell to the recipient one or serve as attachment sites for bacterial viruses and for adherence to mammalian cells.

Cell wall

The vast majority of bacteria have a cell wall containing a special polymer called **peptidoglycan**. The cell wall lies outside the cell membrane, and the rigid peptidoglycan is important in defining the shape of the cell, and giving the cell mechanical strength.

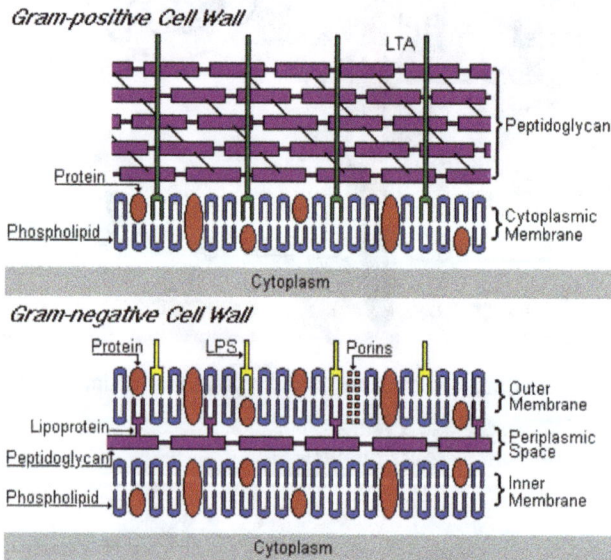

Fig. 3.10. Cell wall of bacteria

The bacterial cell wall is a unique **biopolymer** in that it contains both D- and L-amino acids. Its basic structure is a carbohydrate backbone of alternating units of N-acetyl glucosamine and N-acetyl muramic acid. The NAM residues are cross-linked with oligopeptides. The terminal peptide is D-alanine although other amino acids are present as D- isomers. This is the only biological molecule that contains D-amino acids and it is the target of numerous antibacterial antibiotics. The cell wall of Gram-positive bacteria lies beyond the cell membrane and is largely made up of pepidoglycan. There may be up to 40 layers of this polymer, conferring enormous mechanical strength on the cell wall. Other polymers including teichoic and teichuronic acids also lie in the cell walls of Gram-positive bacteria. These act as surface antigens.

The cell wall of Gram-positive bacteria lies beyond the cell membrane and is largely made up of pepidoglycan. There may be up to 40 layers of this polymer, conferring enormous mechanical strength on the cell wall. Other polymers including teichoic and teichuronic acids also lie in the cell walls of Gram-positive bacteria. These act as surface antigens. In contrast to Gram-positive cells, the cell envelope of Gram-negative bacteria is complex. Above the cell membrane is a periplasm. This area is full of proteins including enzymes. One or two layers of peptidoglycan lie beyond the periplasm. Gram-negative bacteria are thus mechanically much weaker than Gram-positive cells. Beyond the peptidoglycan of the Gram-negative cell wall lies an outer membrane. This has protein channels - porins - through which some molecules may pass easily. The outer side of the Gram-negative outer membrane contains lipopolysaccharide. This provides the antigenic structure of the surface of Gram-negative bacteria and also acts as endotoxin. It is this that is responsible for eliciting the symptoms of Gram-negative shock if it gains access to the bloodstream. Porins and Outer Membrane Proteins (OMPs) act as transporters through the outer membrane.

The various chemicals present in the bacterial cell wall include diaminopimelic acid (DPA), muramic acid and teichoic acid. The other major constituents of bacterial cell wall are amino acids, amino sugars, carbohydrates and lipids. These substances are joined together to form the complex polymeric substance known as peptidoglycan, which is 10-25nm in thickness. Teichoic acid is present only in G+ve bacterial. Petidoglycan is a large polymer composed of three kinds of building blocks;(a) acetoglucosamine (AGA) (b) acetyle muramic acid and (c) a peptide consisting of four or five amino acids of limited variety. Several of them exist in D-configuration.

Peptidoglycan layer is heteropolymer of substituted sugars and amino acids (murein) composed of two acetylated amino sugars-N-acetylglucosamine and N-acetylmuramic acid. Some of the amino acids in murein are unnatural in the sense that they never occur in proteins. Pseudomurein N-acetylomuramic acid in place of N- acetylmuramic acid and unnatural amino acids are not present. Two amino sugars of murein form glycan strands composed of alternating residue of N-acetylglucosamine (G) and N-acetylmuramic acid (M) in beta 1, 4 linkages. Each strand contains 10 to 65 disaccharide units.

Adjacent peptide chains projecting from different glycan strands may be cross linked by formation of peptide bonds between the carboxyl group of terminal D-alanine in one chain and the free á-amino group of meso-diaminopimelic acid in another. Outer membrane is the unit membrane like cell membrane. It is lipid bi-layer containing phospholipids and proteins, and large amount of unique lipid lipo-polysaccharide (LPS) which replaces phospholipids in the outer leaf. One end is hydrophobic inserted in the membrane and hydrophobic on the outer surface. LPS are complex molecules having molecular weight over 10,000 and three distinct regions lipid A, the R core region and O side chain. Lipid A contains 6 or 7 saturated fatty acids attached to a phosphorylated glucosamine dimmer. R core oligosaccharide is a short chain of sugar including two unusual ones, 2-keto-3-deoxyoctonic acid (KDO) and heptose.

Properties associated with bacterial cell walls

Bacteria may be conveniently divided into two further groups, depending upon their ability to retain a crystal violet-iodine dye complex when cells are treated with acetone or alcohol. This reaction is referred to as the *Gram reaction*, named after Christian Gram, who developed the staining protocol in 1884. It may seem a very arbitrary basis, on which to build one's classification system. This reaction, however, reveals fundamental differences in the structure of bacteria. Electron microscopy shows that Gram-negative and Gram-positive bacteria have fundamentally different structures, related to the composition of the cell wall, amongst other things. Cells with many layers of peptidoglycan can retain a crystal violet-iodine complex when treated with acetone. These are called Gram-positive bacteria and appear blue-black or purple when stained using Gram's method. Gram-negative bacteria have only one or two layers of peptidoglycan and can not retain the crystal violet-iodine complex. These need counterstaining with another dye to be seen using Gram's method. A red dye such as dilute carbol fuchsin is often used.

Cytoplasm

It is granular, divisible into two portions; outer cytoplasmic area and inner portion is somewhat denser, chromatin or nuclear area.

Volutin granules

Volutin granules are also known as metachromatic granules. They are prominent in many bacteria, fungi, algae and protozoa. The main constituents in it are meta-phosphate and poly-phosphate. They are highly refractile granules, more prominent as the cell age. They also contain polymerized â-HB (Polyhydroxy butyrate). In some bacteria, polysaccharide granules like starch or glycogen are also found.

Photosynthetic Apparatus

There are three major components of photosynthetic apparatus

(a) Primary light harvesting pigment-antenna

(b) A reaction centre and

(c) Electron transport system.

Carotenoids serve as light harvesting pigments absorbing light in the blue green region. They also play role as quenchers of chlorophyll catalyzed photo-oxidation, thus protecting the photosynthetic apparatus from photo-oxidation. The antenna chlorophylls (bacterio chlorophyll c, d or e) are contained in vesicular sacs, termed *chlorosomes* that are tightly joined to the cell membrane through a base plate. In green bacteria, chlorosomes are surrounded by monolayer membrane which is 4.0 nm thick and composed of lipids and proteins. In blue green bacteria (BGB), unit membrane bound sacs called *thylacoids* are present, which house the reaction centres, electron transport systems and antenna, chlorophyll 'a'. The major antenna pigments, phycobili-proteins are attached to it.

Heterocysts

Heterocysts are heat resistant structures formed during unfavorable conditions in blue green bacteria (BGB). They have normal content of chl 'a', but are devoid of phycobili proteins – a principal antenna pigment of PS II and ribulose bis phosphate carboxylase enzyme, key enzyme of Calvin cycle. Hence they neither fix CO_2 nor produce O_2 in light.

Capsule

Capsule also called slime layer, is in the form of viscous layer or envelope around bacterial cell. It serves several functions as it may provided protective covering to the bacterium or may serve as reservoir of stored food. It may be a site for disposal of waste substances or may also increase infectivity and thus enhancing pathogenicity of the bacterium. The main substances in the capsule layer are polysaccharides like dextran, dextrin, levan and cellulose. In *Bacillus anthracis*, it is a polypeptide, a polymer of D-glutamic acid.

Endospore

Under unfavourable conditions, some Gram positive bacteria have the ability to produce a thick walled oval body (one per cell), called endospore.

These are highly heat resistant, which under favorable conditions, germinate to form vegetative cell *e.g. Bacillus, Clostridium*. Endospores contain large amount of dipicolinic acid, whereas this substance is absent in vegetative cells.

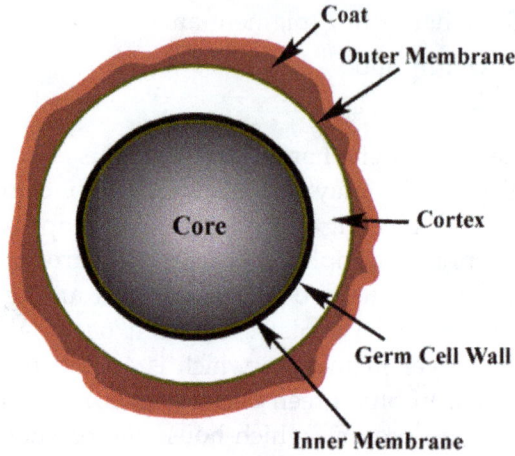

Fig. 3.11. Structure of a bacterial spore

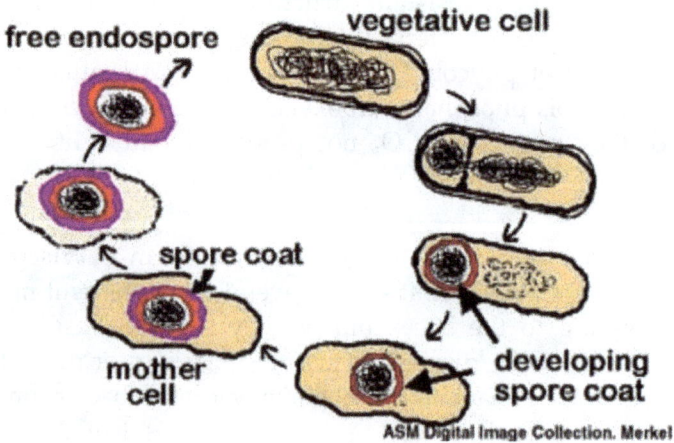

Fig. 3.12. Different stages of spore formation

Mesosome

Mesosomes are membranous intrusions which are involved in reproductive and metabolic processes. It is involved in septum formation during cell division, associated with nuclear matrix and replication. It is also associated with the electron transport chain (ETC) and replication of DNA.

Size of Microbial cell

The dimensions of an average rod shaped prokaryote, the bacterium *Escherichia coli* are about 1x 3 nm. Typical eukaryotic cells may be 2 nm to more than 200 nm in diameter. Cellular metabolic rate and growth rate is inversely proportional to the cell size. The smaller cells have more surface area and thus high metabolic rate and *vice versa*. It is the reason that bacterium plays an important role in the ecosystem.

Cytoplasmic membrane

Cytoplasmic membrane is a thin structure, about 8 nm thick. It is a highly selective barrier, enabling a cell to concentrate specific metabolites and excrete waste materials. It is a phospholipids bi layer, the fatty acids point inward towards each other in a hydrophobic environment and the hydrophilic portions remain exposed to the aqueous external environment.

A few medically important bacteria do not stain easily using conventional stains, and need to be heated to near boiling point in the chosen dye (carbol fuchsin for light microscopy: rhodamine-auramine for fluorescence microscopy) for at least five minutes. This is to allow the dye to penetrate the waxy cell walls. Having taken the stain, these bacteria resist decolorization with both acids and alcohol, and are known as acid-alcohol fast bacteria. This is a property of myco-bacteria. These include *Mycobacterium tuberculosis*, the causal organism of tuberculosis; a chronic infection. Most common is pulmonary tuberculosis, affecting the lung. The kidneys may be infected in renal TB, and there is a rare form of osteomyelitis (bone infection) and meningitis caused by TB. In miliary tuberculosis, the infection is disseminated through the body. Another medically important mycobacterium is *Mycobacterium leprae*, the causal organism of leprosy; a chronic infection of the skin and nerves. Nerve damage leads to a loss of sensation, and ultimately to paralysis. This can lead to tissue damage that can lead to the loss of fingers and toes.

Some bacteria are enclosed within a capsule. This protects the bacterium, even within *phagocytes*, helping to prevent the cell from being killed. Encapsulated bacteria grow as 'smooth' colonies, whereas colonies of bacteria that have lost their capsules appear rough. Rough colonies do not generally cause disease. Encapsulated bacteria do not succumb to intracellular killing as easily as bacteria that lack capsules. Strains of *Streptococcus pnuemoniae* that lack capsules do not cause disease. All the bacteria that cause meningitis are encapsulated. Suspending bacteria in India ink is an easy way of demonstrating capsules. Ink particles cannot penetrate the capsular material and encapsulated cells appear to have a halo around them. This is the Quelling reaction.

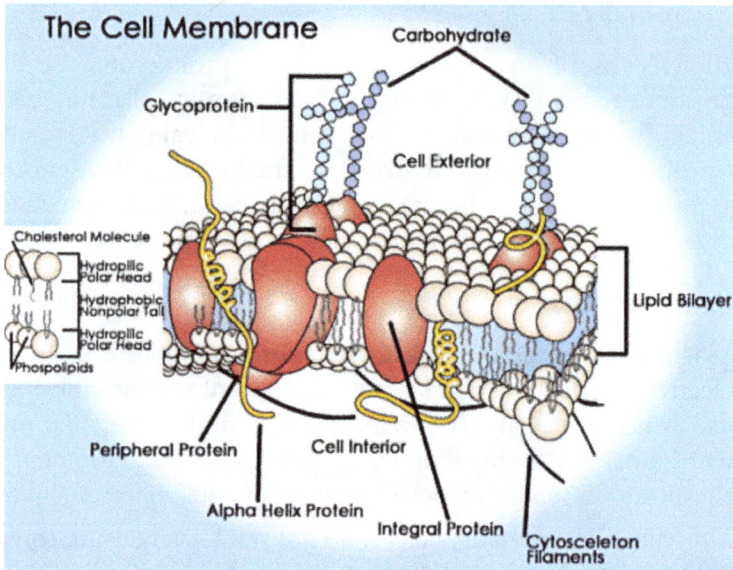

Fig. 3.13: The Cell Membrane

Proteins are embedded in the bi-layer. The overall structure is stabilized by hydrogen bonds and hydrophobic interactions. Cations such as Ca^{+2} and Mg^{+2} also help stabilize the membrane by forming ionic bonds with negative charges of the phospholipids. One major difference in chemical composition of membranes between eukaryotes and prokaryotes is that eukaryotes have sterols in their membranes while sterols are absent from membranes of virtually all prokaryotes. Depending on the cell type, sterols can make up from 5 to 25 per cent of the total lipids of eukaryotic membranes. Sterols are rigid, planar molecules while fatty acids are flexible. Sterols make the CM less flexible and more stable. Membrane rigidity may be necessary in eukaryotes because many of them lack a rigid cell wall and much larger than prokaryotes. Polyenes group of antibiotics (filipin, nystatin and condicidin) react with sterols and destabilize the membrane. Mycoplasmas which lack cell wall require sterol for growth. The sterols become incorporated into the CM and stabilize it. Mycoplasmas are inhibited by polyane antibiotics. Molecules similar to sterols, called hopanoids are present in anaerobic bacteria. The C_{30} hopanoid diploptene is present in hopanoid containing prokaryotes. Unlike the biosynthesis of sterols, hopanoid biosynthesis does not require an oxidation step. Therefore, hopanoids are found in anaerobic prokaryotes.

Archaeal Membranes

In contrast to the lipids in bacteria and eukarya in which ester linkages bond the fatty acids to the glycerol molecule, lipids from Archaea have ether linkages between glycerol and their hydrophobic side chains. In addition, archaeal lipids lack fatty acids and instead have side chains of repeating units of hydrocarbon molecule isoprene. Glycerol diethers and glycerol tetra ethers are the major classes of lipids present in Archaea. They have lipid monolayer instead of a lipid bilayer.

Transport Across Biological membranes

Some small non polar and fat soluble substances such as fatty acids, alcohols and benzene may enter and exit the cell readily by being dissolved in the lipid phase of the membrane. The polar compounds –organic acids, amino acids and inorganic salt do not pass through cell membrane because they are hydrophilic in nature. Hydrophilic substances pass through the membrane through action of membrane transport proteins. There are three classes of transport proteins. Uniporters are proteins that transport a substance from one side of the membrane to the other. Symporters are membrane proteins that carry both substances across the membrane in the same direction. Antiporters transport one substance across the membrane in one direction while transporting the second substance in the opposite direction. Two major mechanisms of energy linked transport are known. Group translocation is the process whereby a substance is transported while simultaneously being chemically modified, generally by phosphorylation. In the process of active transport the substance can accumulate to a high concentration in the cytoplasm in chemically unaltered form. Active transport requires energy in the form of ATP.

A. Group Translocation

Group translocation is a transport process in which the substance is chemically altered in the course of passage across the membrane. Since the product that appears inside the cell is chemically different from the external substrate, no actual concentration gradient of the external solute is produced across the membrane. Examples are glucose, mannose, fructose, N-acetyl glucosamine and β-glucosides, which are phosphorylated during transport by the phosphotransferase system. No energy is involved in group translocation.

B. Active Transport

Active transport is an energy-dependant pumping system, in which the substance being transported combine with membrane-bound carrier, which releases the chemically unchanged substance inside the cell. Substances transported by active transport include some sugars, most amino acids and organic acids, and a number of inorganic ions, such as sulfate, phosphate and potassium.

Table 3.1: Major difference in eukaryotic and prokaryotic cells.

Component	Eukaryotic cell	Prokaryotic cell	Function
Group of organisms where found	Protozoa, algae, fungi, plants and animals	Bacteria and blue green bacteria	-
Cell size	5μm or more in diameter	1-2 μ x 1-4 μ	-
Capsule	Not present (+ in yeast)	Present (Absent in some)	Imparts pathogenicity in bacterial pathogens
Cell membrane	Sterols present	Sterols absent, lipid and proteins are present	Osmotic barrier
Endoplasmic reticulum	Present	Absent	Provide selection to the cells. Ribosomes are attached.
Nucleus	Enclosed in unit membrane; contain chromosomes	No defined nucleus, DNA found dispersed in cytoplasm	Genetic material
Ribosomes	70S in organelle, 80S in cytoplasmic	70S type	Site of protein synthesis
Mitochondria	Present	Absent	Site of respiration
Chloroplast	Present (Absent in yeast and protozoa)	Absent (Blue green bacteria & Green sulfur bacteria contain bacteriochlorophil)	Site of photosynthesis
Golgi bodies	Present	Absent	Site of fat biosynthesis
Lysosomes	Present	Absent	Contains hydrolytic enzymes protect cells from pathogens.
Mesosomes	Absent	Present	Participate in septum formation
Movement	Present (exception in plants)	Present	Locomotion
Locomotic organelle	Protozoans have cilia; Algae have flagella	Bacteria-flagella, cilia	Locomotion
Amoeboid movement	Present in protozoa	Absent	Translocation of solids
Cytoplasmic streaming	Present	Absent	Translocation of solids

Contd...

Table 3.1: _Contd...._

Component	Eukaryotic cell	Prokaryotic cell	Function
Pinnocytosis	Absent (present in Protozoa)	Absent	Intake of liquids
Phagocytosis	Absent (present in Protozoa)	Absent	Intake of solids
Antibiotic sensitivity Penicillin	Resistant	Gram positive bacteria sensitive	Mucopeptidoglycan synthesis is inhibited
Streptomycin	Resistant	Sensitive	Inhibits protein synthesis on 70S ribosome
Cyclohexamide	Sensitive	Resistant	Inhibit protein synthesis on 80S ribosomes
Chloramphenicol	Resistant but sensitive at high concentration	Sensitive	Inhibits protein synthesis on 70S ribosomes
Nystatin	Sensitive	Resistant	Inhibits incorporation of sterols in cell membrane.

Chapter-4

Isolation and Cultivation of Microbes

Growth of microbes under more or less well defined conditions is known as culture. A culture that contains only one kind of microorganisms is known as pure or axenic culture. A culture that contains more than one kind of microorganism is known as mixed culture. If it contains only two kinds of microbes, deliberately maintained in association with one another, it is known as two membered culture. The separation of particular kind of microbe from the mixed population is known as isolation. The growth of microbial populations in artificial environments (culture media) under laboratory conditions is known as cultivation. Microbes do not require much space for cultivation. Artificial environment can be created within a test tube, flask or Petri-dish. The culture vessel should be sterile initially and after the introduction of desired type of microbe, it should be protected from subsequent external contamination. The inoculum (the microbial materials used to seed or inoculate a culture vessel) is introduced on a metal wire or loop sterilized in plugged liquid culture.

Isolation of pure culture by plating methods

For aerobic bacteria, three procedures are followed to form isolate colonies through separation and immobilization of individual organisms on or in a nutrient medium solidified with agar.

(i) Streaked plate method

Series of parallel or non-overlapping streaks are made on the surface of already solidified medium, after proper dilution. The streaked plates are incubated at ambient temperature in BOD incubator.The isolated colonies thus obtained are maintained on slants for further studies.This method is generally used for purification of an isolates from mixture or from contaminants.

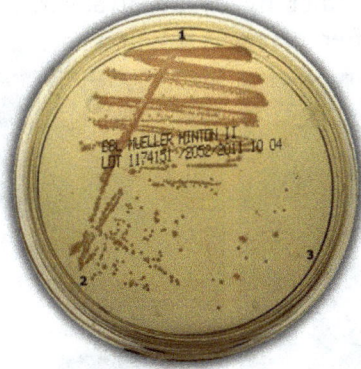

Fig. 4.1. Streaked Plate Method

(ii) Pour plate method

Successive dilution of the inoculums are made and 0.1 ml of proper dilution is put in sterile Petri plates and mixed with the cooled but still molten agar medium which is then allowed to solidify.The plates are incubated for 2-3 days at 28°C depending upon the growth of colonies. The isolated colonies are counted and on this basis we can count number of isolates present in an inoculum.

Fig. 4.2. Pour Plate Method

(iii) Spread plate method

The spread plate method is another technique for counting of microbes in a given sample. In it also, successive dilutions of the given sample are made. First, 10 g of given sample is dissolved in 90 ml of water blank. From it, 1.0 ml is transferred to 9.0 ml of water blank under aseptic conditions. It is mixed well on magnetic stirrer. This process is repeated to obtain proper dilutions. Then 0.1 ml of successive dilutions of the inoculum is spread over already solidified selective medium with the help of spreader. For anaerobic microbes, plates are incubated in a closed container from which air is removed either by chemical absorption or evacuation. The colonies are counted after proper growth.

Fig. 4.4. Spread Plate Method

Sterilization

Sterilization is a treatment that frees the treated object of all living organisms. It can be done by exposure of the object to lethal physical or chemical agents or in the thermo labile state by filtration. When a pure microbial population is exposed to a lethal agent, the kinetics of death are nearly always exponential, the number of survivors decreases geometrically with time. This reflects the fact that all the members of the population are of similar sensitivity. If the logarithm of the number of survivors is plotted as a function of the time of exposure, a straight line is obtained. Its negative slope defines the death rate.

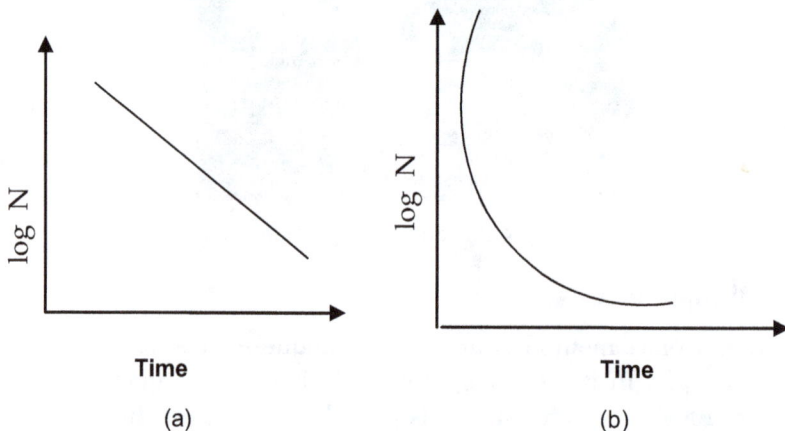

(a)   (b)

Fig. 4.4.

Exponential (logarithmic) order of death of bacteria

(a) The data plotted semi-logarithmically.

(b) The data plotted arithmetically where N is the number of surviving bacteria.

Sterilization by heat

Heat is the most widely used lethal agent for sterilization purpose which is done in two ways

- Dry heating
- Wet heating.

Moist Heat

Heat in the form of saturated steam under pressure is the most practical and dependable agent for sterilization. Steam under pressure provides temperature above boiling point of water. It causes rapid killing of microbes as penetration of heat is more rapid resulting in the coagulation of proteins. It is done with the help of autoclave. The autoclave is operated at a pressure of approximately 15lb/inch^2 (121°C). Wet heating is done in autoclave commonly operated at a steam pressure of 1.06 kg/cm^2 (15lb/ inch^2) above atmospheric pressure, corresponding to a temperature of 121°C. Small volumes of liquid can be sterilized by exposure for 20 minutes; for larger volumes the time of treatment must be extended.

Tyndallization

Some bacteriological media and chemicals cannot be heated above 100°C without being adversely affected. They can be sterilized by fractional sterilization (Tyndallization) by heating the material at 100°C on three successive days with incubation period; on subsequent exposures spores germinate to form vegetative cells to heat the vegetative cells are killed. An apparatus known as the Steam Arnold is used for this technique.

Pasteurization

Milk, cream and certain beverages (beer and wine) are subjected to a controlled heat treatment which kills vegetative stages of microorganisms.

Dry Heating

Hot air sterilization

It is done with the help of hot air oven. For laboratory glassware, two hours exposure to a temperature of 160°C is sufficient for sterilization. Dry heating is done in an atmosphere of air in oven while wet heating is conducted in autoclave by steam. Sterilization by dry heat requires a much greater duration and intensity because heat conduction is less rapid in dry than in moist air. In addition, bacteria can survive in a completely desiccated state as the intrinsic heat resistance of vegetative bacterial cells is greatly increased, almost to the level of spores. Dry heat is used to sterilize glassware or other heat stable solid materials which is generally done at 170°C for 90 minutes in an oven.

Sterilization by chemical treatment

For sterilization of heat labile substances, chemical treatment should be applied. The chemical sterilizing agent should be volatile and toxic, so that it should be easily eliminated from the object after treatment. It is generally done by ethylene oxide, a liquid that boils at 10.7°C. It is generally added to the solutions at a concentration of 0.5 to 1.0 per cent at a temperature of 0 to 4°C. But in aqueous solutions, it is converted into ethylene glycol which is nonvolatile and may have undesirable effects. It is both explosive as well as toxic for human beings and hence, special care must be taken in its handling. It is generally used industrially for sterilization of petri dishes and other plastic objects.

Low Temperatures

Agar-slant cultures of some bacteria, yeasts and molds are stored for prolonged periods of time at refrigerator temperatures of about 4 to 7°C. Many bacteria and viruses can be maintained in deep freeze at –20 to –70°C. Liquid nitrogen at a temperature of –196°C is also used for preservation of semen of animals to be used for race improvement.

Lyophilization

Organisms are subjected to extreme dehydration in the frozen state and then sealed in a vacuum. In the desiccated state, the cultures remain viable for many years.

Radiations

Gamma rays and X-rays which have energies of more than 10 ev, are called ionizing radiations because they have enough energy to pull electrons away from molecules and ionize them. When such radiation passes through cells, it creates free hydrogen and hydroxyl radicals and some peroxides, which in turn can cause different kinds of intra cellular damage. This method is called cold sterilization because ionizing radiations produce relatively little heat in the material being irradiated. It is used to sterilize heat labile materials like food and pharmaceutical articles.

Ultraviolet light

Ultraviolet lamps which emit a high concentration of UV rays in the most effective region, 2600 to 2700°A, are available. Germicidal lamps are widely used to reduce microbial population in hospital operating rooms and aseptic filling rooms in the pharmaceutical industry.

Application of Chemical Agents

Some of the major antimicrobial chemical agents are discussed below

1. Phenol and phenolic compounds.
2. Alcohols
3. Halogens
4. Detergents
5. Gaseous chemo-sterilizers (ethylene oxide, B-propiodactone formaldehyde)

Phenol and Phenolic Compounds

Phenol (carbolic acid) and its derivatives probably act primarily by denaturing cell proteins and damaging all membranes. Hexyl-resorcinol greatly reduces surface tension and thus contributes to its antimicrobial action. Phenolic substances may be either bactericidal or bacteriostatic depending upon the concentration used. Aqueous solutions of phenol from 2 to 5 per cent are generally used to disinfect various materials like sputum, urine, feces and utensils. The cresols are several times more germicidal than phenol.

Hexyl-resorcinol, another derivative of phenol, is marketed in a solution of glycerin and water. It is a strong surface-tension reductant, which may contribute to its high bactericidal activity.

Alcohols

Ethyl alcohol, CH_3CH_2OH in concentrations ranging from 50 to 70 per cent is effective against vegetative cells. Generally, 70 per cent alcohol is used for bactericidal activity. Methyl alcohol is less bactericidal than ethyl alcohol and is highly poisonous also. The higher alcohols – propyl, butyl, amyl are more germicidal than ethyl alcohol. However, alcohols with molecular weight not higher than that of propyl alcohol are not miscible in all proportions with water. Therefore, they are not commonly used as disinfectants. Alcohols are protein denaturants, solubilize lipids and thus damage cell membrane. They are also dehydrating in nature and thus bactericidal as well as bacteriostatic. The high concentrations of alcohol remove so much water from the cell that the alcohol is unable to penetrate, due to which 70 per cent alcohol is most appropriate.

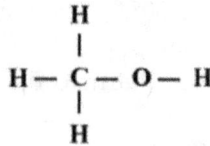

$$H - \underset{\underset{H}{|}}{\overset{\overset{H}{|}}{C}} - O - H$$

Methyl alcohol
CH_3OH

Halogens

Pure iodine is a bluish-black crystalline element having a metallic luster. It is slightly soluble in water but readily soluble in alcohol. It is used as germicidal agent in the form of tincture of iodine. It is effective against all kinds of bacteria, also possesses sporicidal activity. Iodine solutions are chiefly used for the disinfection of skin. It may cause halogenations of tyrosine units of enzymes and other proteins. It also serves as oxidizing agent. The compressed chlorine gas in liquid form is universally employed for purification of municipal water supplies. It is difficult to handle in elemental form and hence in the compound form it is equally effective. Calcium hypochlorite, $Ca (OCl)_2$ also known as chlorinated lime and sodium hypochlorite (NaOCl) are popularly used domestically and industrially. Solutions of sodium hypochlorite of 1 per cent concentration are used for personal hygiene and as a household disinfectant. The germicidal action of chlorine and its compounds comes through the hypochlorous acid formed when free chlorine is added to water:

$$Cl_2 + H_2O \longrightarrow HCl + HClO \longrightarrow \text{Hypochlorous acid.}$$

The hypochlorous acid releases nascent oxygen which is a strong oxidizing agent.

Gaseous Chemo-sterilizers

Plastic heat sensitive items such as syringes, laboratory ware etc. are sterilized by gases. The main agents include ethylene oxide, B-propiolacton and formaldehyde. Ethylene oxide is liquid at temperature below 10.8°C and above this it vaporizes rapidly.

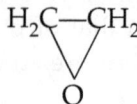

$$\underset{O}{\overset{H_2C - CH_2}{\bigtriangledown}}$$

It is used for sterilizing heat labile materials. It is highly sporicidal. It has high penetration and has broad spectrum of activity against microorganisms. It is effective even at low temperature and does not damage the materials exposed to it. Formaldehyde is marketed in aqueous solution as formalin which contains 37 to 40 per cent formaldehyde. For sterilization with formaldehyde, the temperature should be around 22°C and relative humidity between 60 to 80 per cent. But it has limited ability to penetrate covered surfaces.

Sterilization by Filtration

The principal laboratory method used to sterilize solutions of heat labile materials is filtration through millipore filters which are capable of retaining microorganisms. Some substances like enzymes, vitamins, antibiotics are thermo labile. They are sterilized by use of bacteriological filters which include asbestos pad in the Seitz filter, porcelain, sintered glass and millipore filter. Porosity is not the single criteria for these filters. Other factors such as electric charge of the filter, electric charge carried by the organisms and nature of the fluid being filtered, have a bearing on the efficiency of filters. Membrane or molecular filters are composed of biologically inert cellulose esters. They have porosity in the range of 0.01 to 10 μm.

Control of pH in the medium

To prevent excessive changes in hydrogen ion concentration, either buffers or insoluble carbonates are often added to the medium. The phosphate buffers which consist of mixtures of mono-hydrogen and di-hydrogen phosphates e.g. K_2HPO_4 and KH_2PO_4 are the most useful ones. KH_2PO_4 is a weakly acidic salt whereas K_2HPO_4 is slightly basic, so that an equimolar solution of the two is very nearly neutral, having a pH of 6.8. If a limited amount of strong acid is added to such a solution, part of the basic salt is converted to the weakly acidic one.

$$K_2HPO_4 + HCl \longrightarrow KH_2PO_4 + KCl$$

If, however, a strong base is added, the opposite conversion occurs

$$KH_2PO_4 + KOH \longrightarrow K_2HPO_4 + H_2O$$

Generally, about 5.0 g of potassium phosphates per liter of medium can be tolerated by bacteria and fungi. When large amount of acid is produced by a culture, carbonates may be added to media as "reserve alkali" to neutralize the acids as they are formed. In the presence of hydrogen ions, carbonate is transformed to bicarbonate and bicarbonate is converted further to carbonic acid, which decomposes spontaneously to CO_2 and water. This sequence of reactions, all of which are freely reversible, can be summarized

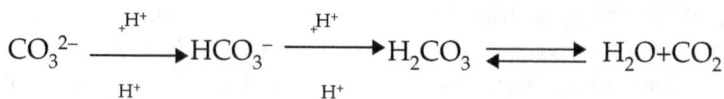

$$CO_3^{2-} \xrightarrow[\text{-H}^+]{\text{+H}^+} HCO_3^{-} \xrightarrow[\text{-H}^+]{\text{+H}^+} H_2CO_3 \rightleftharpoons H_2O + CO_2$$

The soluble carbonates, such as Na_2CO_3, are strongly alkaline and are therefore, not suitable for use in culture media. Of insoluble carbonates, finely powdered chalk ($CaCO_3$) is the most generally employed. If the medium has high phosphate concentration, on sterilization, the formation of a precipitate occurs. To avoid this problem a chelating agent is generally used. The most common chelating agent is ethylene di-amine-tetra acetic acid (EDTA), at a concentration of approx 0.01 per cent.

MICROSCOPE AND MICROSCOPY

A microscope (from the *Ancient Greek*: *mikrós*, "small" and *skopeîn*, "to look" or "see") is an *instrument* used to see objects that are too small for the naked eye. The science of investigating small objects using such an instrument is called *microscopy*. *Microscopic* means invisible to the eye unless aided by a microscope.

Microscopic examination of microorganisms makes use of either the light microscope or the electron microscope. All microscopes employ the principle that specific lenses magnify the image of such microorganism that details of its structure are most apparent. In addition to magnification, however, is resolution, the ability to distinguish two adjacent points as separate. Although magnification can be increased virtually without limit, resolution cannot. Resolution is dictated by the physical properties of light. For light microscope, the limits of resolution are about 0.2 nm (200 nanometers).

Compound Light Microscope

Several types of light microscopes are commonly used in microbiology as bright field, phase contrast, dark field and fluorescence. The bright field microscope is most commonly used which comprises of two series of lens (objective lens and ocular lens) which function together to resolve the image.

Magnification and Resolution of a Compound Light Microscope

The total magnification of a compound microscope is the product of the magnification of its objective and ocular lenses. Resolving power is a function of the wavelength of light used and an innate property of the objective lens known as its nuclear aperture (a measure of light gathering ability). Lenses with higher magnification usually have higher numerical apertures. The highest resolution possible in a compound light microscope is about 0.2 nm. This means that two objects closer together than 0.2 nm are not resolvable as distinct and separate. Lenses on which oil is to be used are called Oil-immersion lenses. Oil is used with these lenses because it has a higher numerical aperture than air (which has a numerical aperture of 1) and also has higher refractive index. This greatly improves resolution.

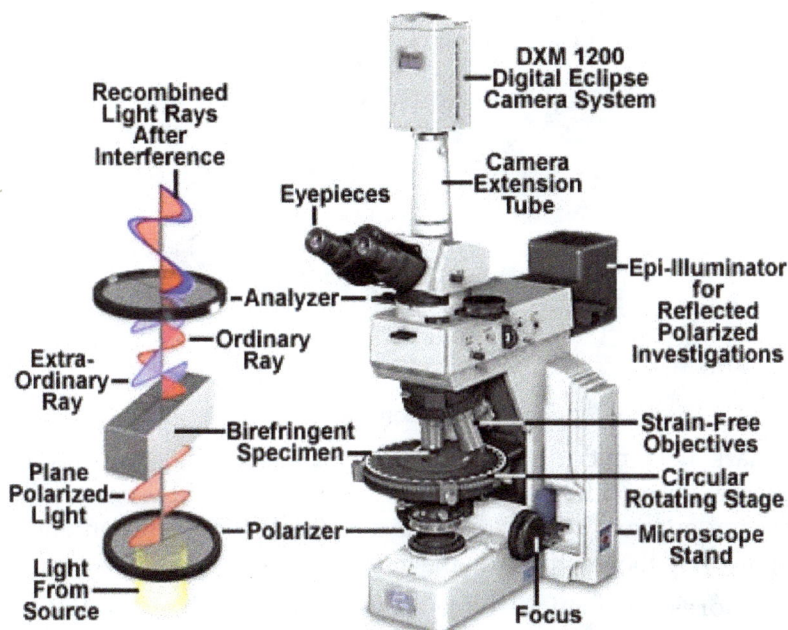

Fig. 4.5. Polarized Light Microscope Configuration

Phase contrast, Dark Field and Fluorescence microscope

The phase contrast microscope improves contrast differences between cells and surrounding medium, making it possible to see cells without staining them. The basic principle of it is that the cells differ in trefractive index with the surroundings and hence bend some of the light rays that pass through them. Light passing through a specimen of refractive index different from that of the surrounding medium is retarded. This effect is amplified by a special ring in the objective lens of a phase contrast microscope, leading to the formation of a dark image on a light background. In phase contrast microscope, web amount of microorganisms can be easily observed without staining. Staining generally kills the micro-organisms and thus distorts their features.

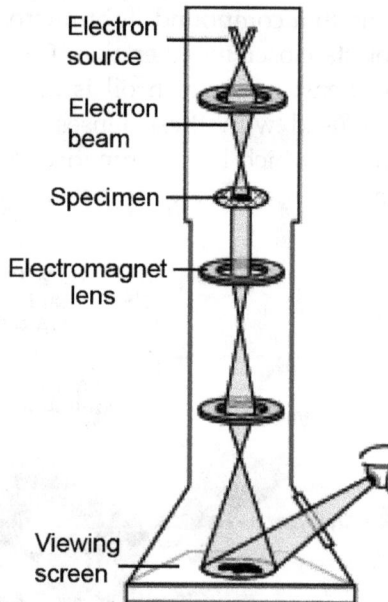

Fig. 4.6. Phase Contrast Microscope

The dark field microscope is a light microscope in which the lighting system has been modified to reach the specimen from the sides only. In it, the specimen appears bright against a dark background. Resolution in it is very high, even the motility of the microorganism can be observed.

The fluorescence microscope is used to visualize specimens that fluorescence. They emit light of one color and absorb light of another color. The cells florescence either due to presence of natural fluorescent substance or the cells have been treated with a fluorescent dye.

Staining

Dyes can be used to stain cells and increase their contrast in bright field microscope. They are organic compounds and each class of dye has an affinity for specific cellular materials. Many dyes are positively charged (cationic) and combine strongly with negatively charged cellular constituents such as nucleic acids and acidic polysaccharides. Examples of cationic dyes include methylene blue, crystal violet and safranin.

Electron Microscope

To study the internal structure of cells, a transmission electron microscope (TEM) is essential. In the TEM, electrons are used instead of light rays and electromagnets function as lenses, the whole system operating in a high vacuum. Electron beams do not penetrate very well hence a single bacterial cell is cut into thin slices, which are then individually examined with electron microscope. Special electron microscope stains such as osmic acid, permanganate, uranium, lanthanum or lead are used which are composed of atoms of high electronic weight. They scatter electrons well and thus improve contrast. All electron microscopes are fitted with cameras to allow a photograph, called an electron micrograph, to be taken.

Chapter 5

Replication, Transcription and Translation in Bacteria

Replication refers to the copying/synthesis of new DNA through various enzymes such as DNA polymerase and ligase. Transcription is the synthesis of mRNA from a DNA template. Translation is the synthesis of polypeptides using mRNA as a guideline in ribosomes, with tRNA as the carrier of amino acids. Replication, transcription, and translation are the same processes in bacteria, eukarya and archea (they accomplish the same thing). The mechanisms and enzymes between them are different however.

Replication

Replication may be uni-directional or bi-directional. It involves following three steps:

1. Initiation: The point at which the replication occurs is called the replication fork or growing point. Enzymes that are able to synthesize new DNA strands on a template strand are called DNA polymerases. Both, prokaryotic and eukaryotic cells contain more than one DNA polymerase. Only one DNA polymerase provides the replicase function. Others are involved in repair or subsidiary functions. In eukaryotic cells DNA polymerase α level increases during S phase of cell division, DNA polymerase β causes repair mechanism and DNA polymerase γ synthesizes mitochondrial DNA. In prokaryotic cells, Pol I causes the repair of damaged DNA and subsidiary role in semi-conservative replication, function of DNA Pol II is unknown and DNA Pol III, multi subunit protein is responsible for *de novo* synthesis of DNA. Each can extend a DNA chain by adding nucleotides one at a time to 3' end. Synthesis occurs in 5'-3' direction. All the bacterial enzymes possess 3'-5' exo-nucleolytic activity for proof reading.

2. Elongation: DNA synthesis is semi discontinuous. As the replication fork advances, daughter strands are synthesized on both of the exposed parental single strands. On one strand, DNA synthesis proceeds continuously in the 5'-3' direction as the parental duplex unwound. This is called the leading strand. On the other hand a series, of small fragments are synthesized, each 5' to 3', and then they are joined together to create an intact lagging strand; hence it is called discontinuous replication. These fragments are called okazaki fragments. Each okazaki fragment start with a primer -10 base long, a short sequence that provide the 3'-OH end for extension by DNA polymerase. The 5'-3' exo-nuclease activity to separate primer and then sealing the nick by DNA ligase. Sequences prior to the start point are called as upstream; those after the start point are called downstream. Consensus sequence in *E. coli* is TATAAT or Pribnow box while in eukaryotes, it is 10 base sequence called as TATA or Hogness Box.

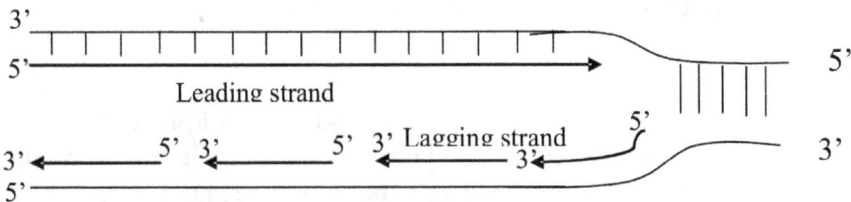

Fig. 5.1. DNA Synthesis – semi discontinuous

3. Termination

Eukaryotes initiate DNA replication at multiple points in the chromosome, so replication forks meet and terminate at many points in the chromosome; these are not known to be regulated in any particular way. Because eukaryotes have linear chromosomes, DNA replication is unable to reach the very end of the chromosomes, but ends at the *telomere* region of repetitive DNA close to the end. This shortens the telomere of the daughter DNA strand. This is a normal process in *somatic cells*. As a result, cells can only divide a certain number of times before the DNA loss prevents further division. (This is known as the *Hayflick limit*.) Within the *germ cell* line, which passes DNA to the next generation, *telomerase* extends the repetitive sequences of the telomere region to prevent degradation. Because bacteria have circular chromosomes, termination of replication occurs when the two replication forks meet each other on the opposite end of the parental chromosome. *E coli* regulate this process through the use of termination sequences that, when bound by the *Tus protein*, enable only one direction of replication fork to pass through. As a result, the replication forks are constrained to always meet within the termination region of the chromosome. It is of two types' that is the core enzyme can terminate *in vitro* at certain sites in the absence of any other factor rho independent or simple terminator. Terminator sequences include palindromic regions that form hair pins varying in length from 7 to 20 bp. The stem loop structures are followed by a run of U residues in rho-independent but not in rho-dependent sites.

TRANSCRIPTION

It mainly involves three steps initiation, elongation and termination.

1. Initiation

Initial recognition of duplex DNA, unwinding to generate a short single stranded region and incorporation of the first nucleotide of the RNA chain is called initiation. The entire sequence of DNA that is necessary for these reactions is called the promoter. The site at which the first nucleotide is incorporated is called the start point or start site.

2. Elongation

It starts when a few bases are incorporated into an RNA chain, forming an RNA-DNA hybrid. To continue synthesis, the enzyme moves along the DNA, unwinding the double helix to expose new segment of the template in single stranded condition. Elongation involves the movement along DNA of short segment that is transiently unwound, existing as a hybrid RNA-DNA duplex and a displaced single strand of DNA. The complete enzyme or holoenzyme subunit constitution is $1\sigma\alpha_2\ \beta\beta$ while the core enzyme subunit constitution is $\alpha_2,1\beta,1\beta.$' Only the holoenzyme can initiate transcription, but then the sigma factor is released, leaving the core enzyme to undertake elongation. Thus the core enzyme has the ability to synthesize RNA on a DNA template, but cannot initiate transcription at the proper sites. Eukaryotic RNA polymerase complex occupies three different locations each with a large number of polypeptide subunits. RNA Pol I is present in nucleus transcribing the gene coding for rRNA. It accounts for 50-70 per cent of cellular RNA synthesis. RNA Pol II occurs in nucleoplasm and responsible for 20-40 per cent of cellular activity, for synthesizing heterogeneous nuclear hnRNA (the precursor of mRNA). RNA Pol III accounts for 10 per cent activity of the nucleoplasm, as responsible for tRNAs synthesis.

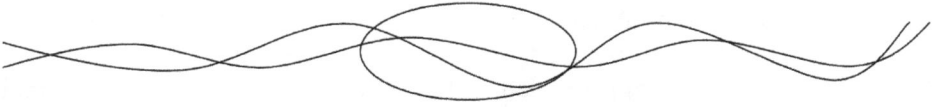

Fig. 5.2. INITIATION – RNA Polymerase binds to duplex DNA

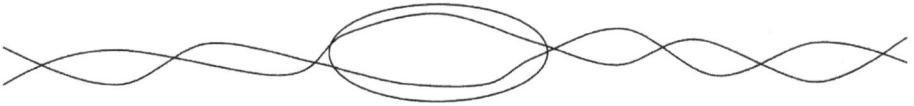

Fig. 5.3. RNA SYNETHESIS – Unwinding of DNA

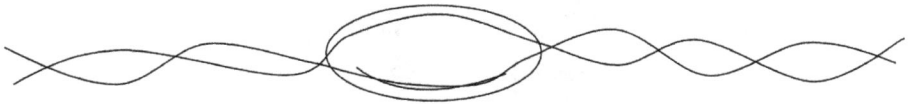

Fig. 5.4. RNA SYNETHESIS by base pairing with one strand of DNA

3. Termination: The ability to recognize the point at which no further bases should be added to the chain. When the last base is added to the RNA chain, the RNA-DNA hybrid is disrupted and DNA reforms in duplex state. Both the enzyme and RNA are released from it. The sequence of DNA required for these reactions is called the terminator.

PROTEIN BIOSYNTHESIS

In bacteria, about 15 amino acids are added to a growing polypeptide chain every second. Seventy S ribosome dissociates into 30S and 50S subunits; Mg^{+2} ion concentration is reduced. The only site that can be entered by an incoming aminoacyl- t RNA is the A site or entry site. The codon representing the last amino acid to have been added to the chain lies in the P site or donar site. First initiation codon was found to be AUG, in few cases it is GUG which is called initiation triplet. A site on the mRNA that precedes the initiator codon by approximately 10 nucleotide, is called as ribosomal binding site or a Shine Delgarno sequences. It is complementary to the 3' hydroxyl end of the 16sRNA component of 30S ribosomal subunit.

1. Initiation

The reactions that precede formation of peptide bond between the first two amino acids of protein. Initiation of polymerization is preceded by the formation of 30S subunit IF_3 and GTP complex.

Protein synthesis

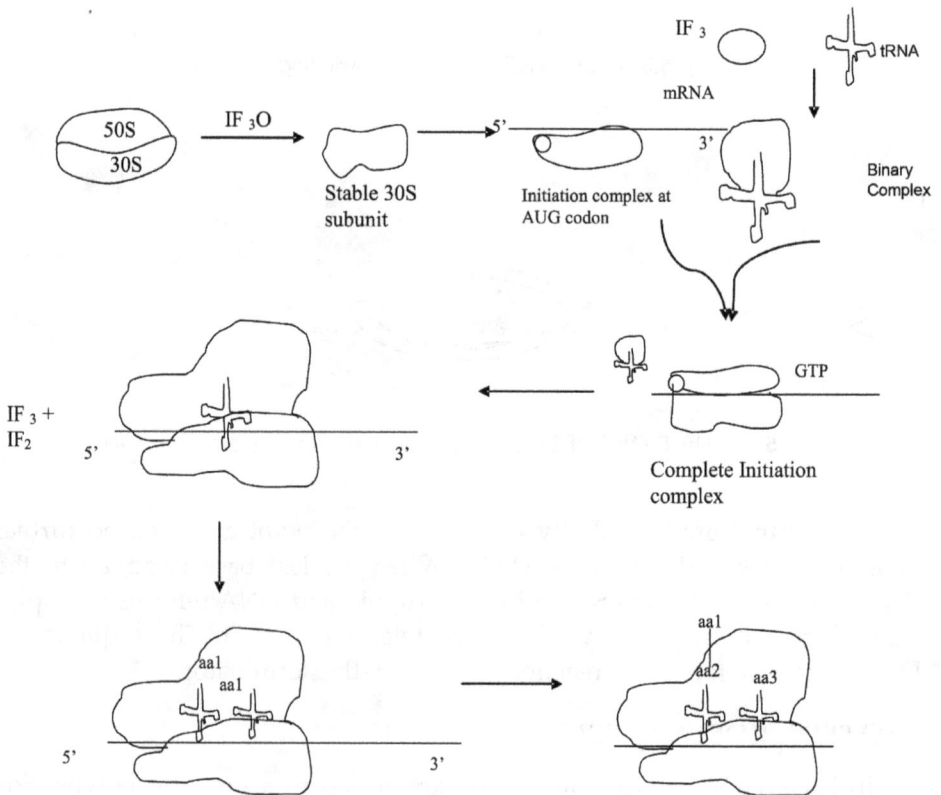

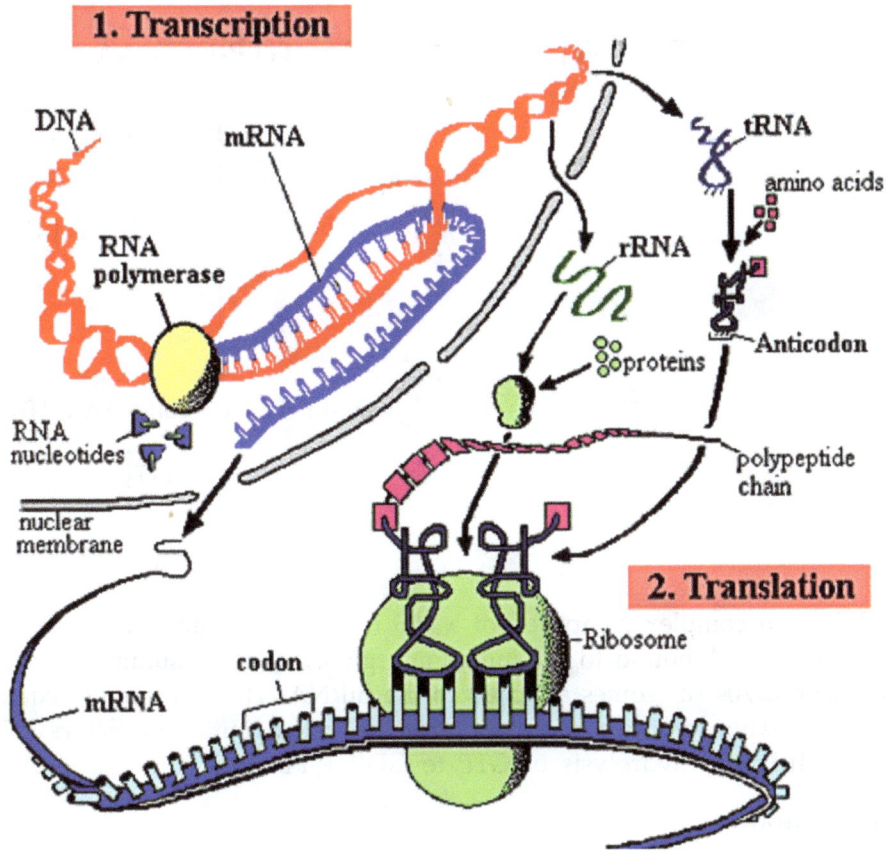

1. Transcription

DNA

mRNA

RNA polymerase

RNA nucleotides

nuclear membrane

tRNA

amino acids

rRNA

Anticodon

proteins

polypeptide chain

2. Translation

Ribosome

codon

mRNA

Fig. 5.5. Protein synthesis

Termination Codons

- UAA – Ochre Codon
- UAG – Amber Codon
- UGA – Opal Codon

Eukaryotes – Nine initiation factors

- Elongation factors – eEF_1 and eEF_2
- Release factors – RF_1 – UAA and UAG

RF_2 – UAA and UGA

RF_3 – stimulates others

$$\text{EFT}_4 \xrightarrow{\text{GTP}} \textbf{GTP-EF-T}_4 \xrightarrow{\text{AA-tRNA}} \textbf{GTP-EFT}_4\textbf{-AA-tRNA}$$

$$\downarrow \text{Ribosome}$$

$$\text{Ribosome-GTP-EF-T}_4\text{-AA-tRNA}$$

$$\downarrow$$

$$\text{GDP-EF-T}_4 + \text{Ribosome-AA-tRNA}$$

$$\downarrow \text{EF-T}_5$$

$$\text{GDP} + \text{EFT}_4 + \text{EFT}_5$$

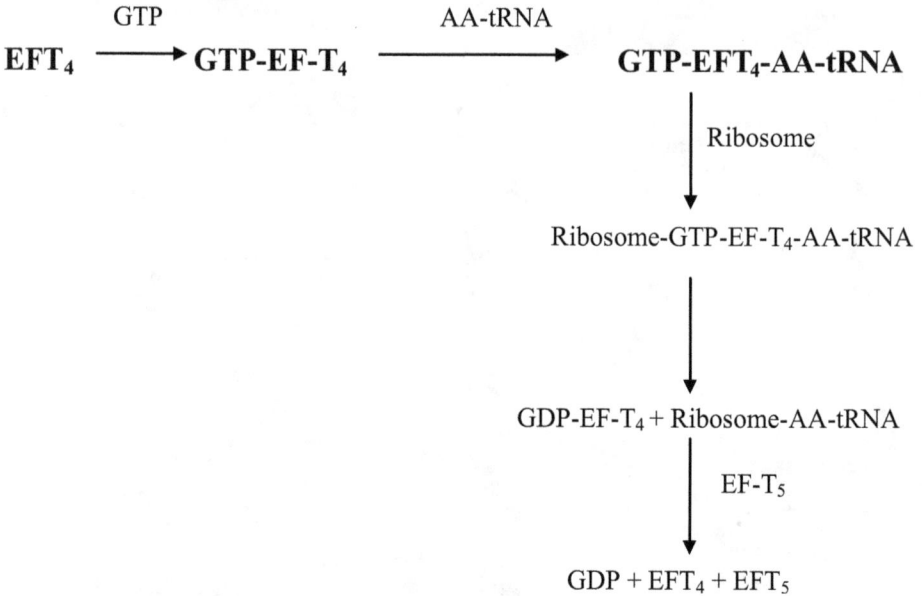

Initiation complex composed of a 30S ribosomal subunit and a formyl methionlyl tRNA bound to the initiation region. Then 50S subunit attaches producing a 70s ribosomes attached to the mRNA. These reactions require the participation of three accessory proteins, called initiation factors (IF1, IF2 and IF3) and hydrolysis of GTP to GDP + Pi.

2. Elongation

It includes all reactions for synthesis of first peptide bond to the addition of last amino acid to the chain. Amino acids are added one at a time, and the addition of amino acids is the most rapid step in the protein synthesis. In elongation of peptide chain 70S ribosomes have two sites, designated as A (aminoacyl) and P (peptide) sites, that binds tRNAs and attached amino acids or peptides. The initiation complex is formed with formylmethionyl tRNA and P site; all other aminoacyl tRNAs enter the A site. It involves three steps- recognition, peptidyl transfer and translocation.

(a) Recognition: A molecule of aminoacyl tRNA attaches to the A site with a sequence of three bases (a codon) on the mRNA molecule. Two protein factors, called elongation factors (EF Tu and EF Ts) and one GTP molecule are involved in recognition.

(b) Peptidyl transfer: After recognition, both ribosomal binding sites are occupied by aminoacylated tRNAs. The peptide bond is formed between the terminal carboxyl group of the peptide and amino group of the amino acid in the A site. It does not require accessory protein factors and additional energy.

(c) **Translocation**: The free tRNA molecule in the P site is released and ribosome moves three base down the mRNA, thereby moving the peptide bearing tRNA form the A site to the P site and having empty A site for the next codon to be read. It requires one accessory protein EF G and one GTP molecule.

3. Termination

Steps that are needed to release the completed polypeptide chain and dissociating ribosome from the mRNA are called termination. Successive amino acids are added to the peptide chain in the order encoded by the sequence of codons in the mRNA molecule till nonsense codon (UAG, UGA and UAA) is reached which causes the releases of the completed protein from the 70S ribosomes one RF release factor is also involved.

Chapter-6

Microbial Growth and Reproduction

Growth: Growth for microorganism is an increase in cell number.

Generation time: It is the time required for a population of microbial cells to double the original number and is also called doubling time.

Batch culture: A closed system of having microbial culture of fixed volume.

Chemostat: A device that allows for continuous culture of microorganisms in which both growth rate and cell number can be controlled independently.

Exponential growth: Growth of microorganisms where the cell number double within a fixed time period. This type of increase in cell numbers is called as logarithmic increase or exponential increase.

Growth rate: Change in cell number or cell mass per unit time.

Binary fission: Bacteria are unicellular and divide asexually into two independent cells, known as binary fission. The time required for a complete growth cycle in bacteria in variable and is dependent on a number of factors, both nutritional and genetic. Generation time for *E. coli* is 20 minute.

Growth can be defined as the increase in number of cells in relation to time. Since bacteria divide by binary fission, always the number is doubled. Therefore, the time taken to double the cell number of cells is called the generation time or doubling time.

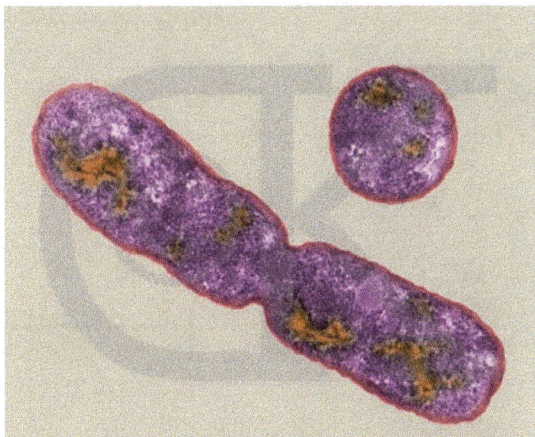

Fig. 6.1. Division in bacterial cell

Doubling or generation time is different from bacteria to bacteria e.g. in *E. coli* it is 20 minutes while for *Rhizobium,* it is 400 minute. Growth can also be measured as an increase in microbial mass. The interval for the formation of the two cells from the one is called one generation and time required for this to occur is generation time or doubling time. Cell number is represented logarithmically and time scale is represented arithmetically (semi-logarithmic graph) resulting in straight line. Growth curve can be represented as follows

$$N = N_0\ 2^n.$$

Where, N is final cell number, N_0 is the initial cell number and n is number of generations that have occurred during the period of exponential growth. The generation time g of the cell population can be calculated as t/n where t is simply the hours or minutes of exponential growth.

$$N = N_0\ 2^n$$

$$\log N = \log N_0 + n \log 2$$

$$n \log 2 = \log N - \log N_0$$

$$n = \frac{\log N - \log N_0}{\log 2} = \frac{\log N - \log N_0}{0.301}$$

Growth curve

When bacteria grow, they show 3-4 distinct phases (stages) of growth, classified as follows:

a. Lag phase   b. Log phase or exponential phase

c. Stationary phase   d. Death phase.

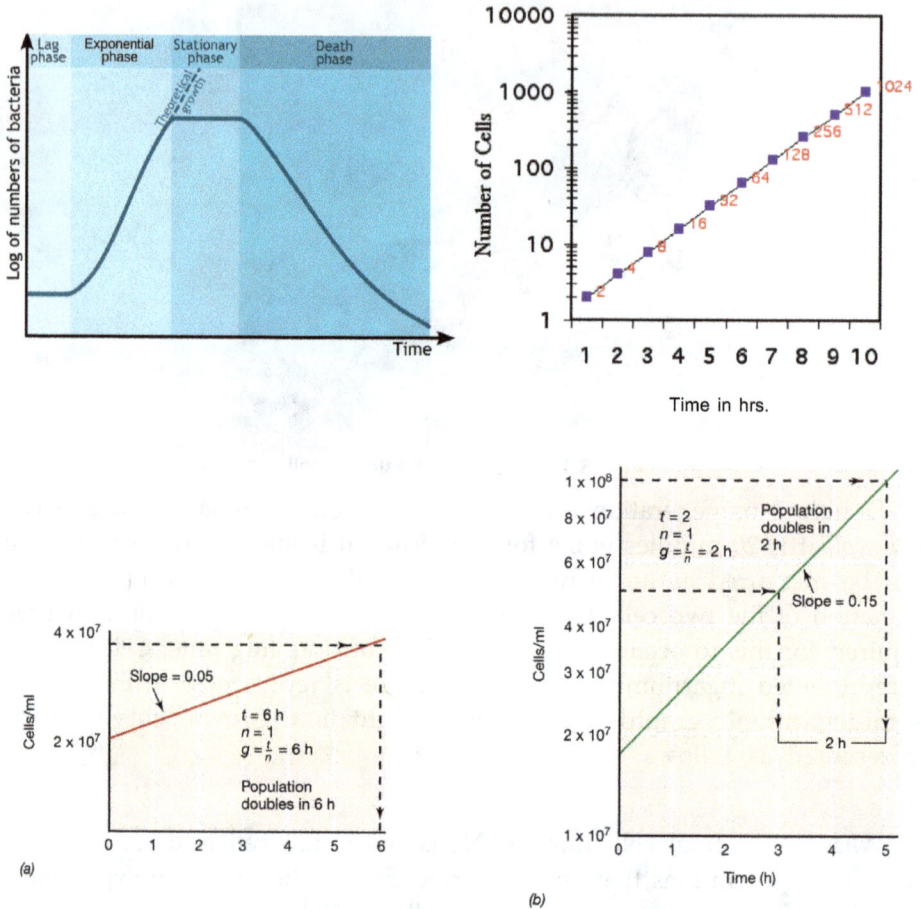

Fig. 6.2. Growth curve

A. Lag phase: There is no increase in cell number during this stage. In this stage the cell prepares itself for division. There is initiation of replication and synthesis of enzymes and structural proteins.

B. Log phase: During this phase, cell number increases exponentially. Here the increase in cell number is maximum. Length of this stage depends on the composition of the medium and environmental conditions.

C. Stationary phase: During this phase, there is no net increase in cell number. The total number of viable cells remains constant as the number of the cells increase and equals to the number of cells that die.

D. Death phase or Decline phase: During this phase, the death of cells is exponential. Therefore, number of cells decrease exponentially.

How to measure growth fungi

The growth in fungi can be determined by determining increase in dry weight or wet weight.

Synchronous growth

Microbial population can be maintained in a state of exponential growth over a long period of time by using a system of continuous culture. The cell concentration in a continuous culture system remains constant. Continuous culture system can be operated as chemostat or turbidostat. In the chemostat, the flow rate is set at a particular value and rate of growth of culture adjusts to this flow rate. In turbidostat, the system includes an optical sensing device which measures the absorbance of the culture.

Quantitative measurement of bacterial growth

Various methods includes

1. Total count (directly by microscopy or an electronic particle counter).

Limitations of Direct count by microscope
- Dead cells are also counted
- Small cells are difficult to see under the microscope
- Phase contrast microscope is required when the sample is not stained.

2. Cell mass (directly by weighing or a measurement of cell nitrogen or indirectly by turbidity).

3. Cell activity (indirectly by relating the degree of biochemical activity to the size of the population).

4. Viable count: Also called as plate count. There are two ways of performing plate count
- Pour plate method
- Spread plate method.

Bacteria can be counted easily and accurately with the Pteroff Hausser counting chamber.

SEXUAL REPRODUCTION IN BACTERIA

It is also known as sex duction. It is of three types:

a. Conjugation

b. Transformation.

c. Transduction

a. Conjugation

It is the unidirectional transfer of the genetic material (DNA) by one bacterial cell to the other. It requires the establishment of physical contact between the donor and the recipient cells. The physical contact between the mating bacteria is established through the F fertility pilus. The bacterial strains that have F pili act as donor; designated as F⁺ strains. The strains lacking F pili are designated as F⁻ strains (recipient or female cell). The precise portion of the DNA that is transferred depends upon the time of mating. Conjugative plasmids, typified by the F plasmid of *E. coli* have the ability to transfer themselves between the cells. The *tra* gene carried by the F plasmid contains all the information for the conjugative process. The plasmid is nicked in one strand at a site called ori T, the origin of transfer. The intact strand acts as a template for DNA synthesis, while the nicked strand is displaced and transferred to the recipient cell. Strains which have an F plasmid integrated into the chromosome are called HFr (high frequency recombination) strains. Sex pilus form the conjugation tube between the two bacterial cells. Only a part of genetic material from the donor cell is transformed to the recipient one, forming merozygote. Conjugation was first discovered by Lederberg and Tatum in 1964.

Merozygote contains the complete genetic complement of the recipient (endozygote) and only a portion of the genetic complement of the donor (exogenote).

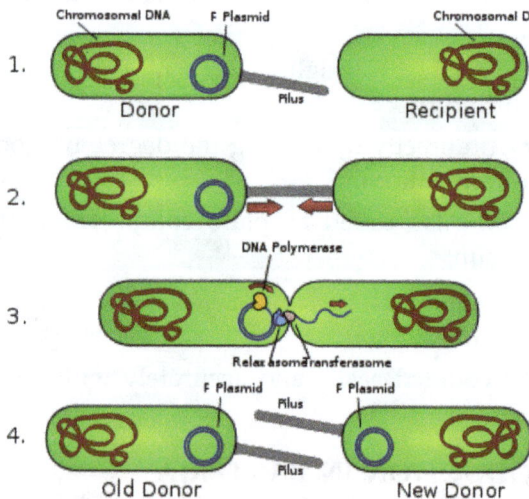

Fig. 6.3. Conjugation

(b) Transformation

The DNA is released from the bacterial cell in surrounding medium and the recipient cell incorporates it into itself from the medium. It was discovered by the Griffith in 1928 in *Streptococcus pneumoniae*. It is of two types

- **Natural Transformation**: Many bacteria which become competent under ordinary conditions of the culture growth.

- **Artificial transduction**: Such bacteria which are not competent under ordinary conditions, but are made competent by a variety of highly artificial treatments such as exposure of the cell to high concentration of divalent ions etc.

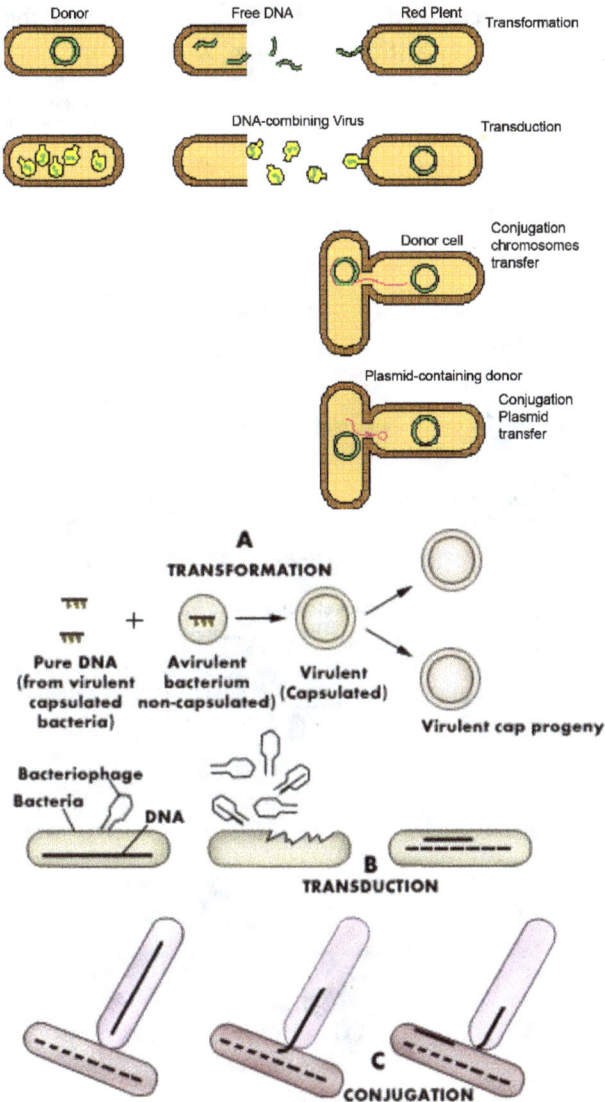

Fig. 6.4. Transformation

c. Transduction

The DNA is transferred from one bacterial cell to another as a result of the formation of an aberrant phage in which some or its entire normal competent DNA is replaced by bacterial DNA (Donor DNA). When such a phage virion attaches to and introduces this DNA into another bacterial cell, genetic exchange occurs. It is of two types

- **Generalized or non-specialized transduction:** In this phage it mediates the exchange of any two bacterial gene e.g. P_{22} in *Salmonella*.

- **Restricted or specialized transduction:** It mediates the exchange of limited number of specific genes. e.g. λ phage in *E. coli*.

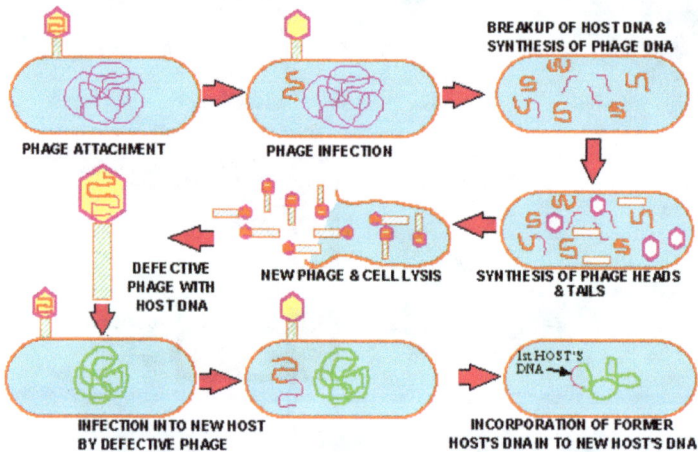

Fig. 6.5. λ phage in E. coli.

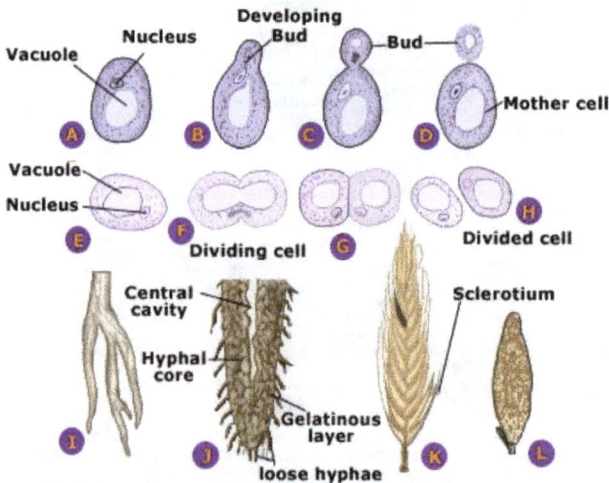

Fig. 6.6. Modes of vegetative reproduction in fungi.
A-D. Budding; E-H. Fission; I-J. Rhizomorph; K-L Scterotia.

Reproduction in fungi

Reproduction in case of fungi occurs either by budding as in yeast and by elongation of apical cells as in filamentous fungi and thus forming asexual spores. These asexual spores germinate to form hyphe under favorable conditions. Sexual reproduction also occur in fungi. In budding there is no specific part where buds arise.

Chapter 7

Viruses

Virus is a Latin word named after disease causing substances; also called poison. Viruses can be defined as a large group of disease causing agents that are composed of infectious nucleic acids encapsulated in a protein coat. They may possess membranes but do not have cytoplasm or metabolism of their own.

Discoveries

- Working on rabies, Louis Pasteur in 1884 discovered that viral diseases could be transmitted from one host to another host under laboratory conditions.
- Mayer (1886) discovered mosaic disease of tobacco plants.
- D. Iwanowsky (1892) discovered that viruses can pass through bacterial proof filters.
- Loffeur and Frosch (1898) discovered FMV (foot and mouth viruses) in cattles.
- In 1915, F.W. Twort and in 1917, F. d'Herelle discovered bacteriophages.

Properties of viruses

- Viruses are obligate parasites. They cannot be grown on nonliving media. They need a living host for its survival and multiplication.
- Viruses are ultramicroscopic. They cannot be seen by light microscope; can be visualized only by electron microscope.
- Viruses are composed of proteins and nucleic acids. Nucleic acid can

be DNA or RNA. DNA and RNA may be single stranded or double stranded. Plant viruses contain only RNA, while animal viruses contain RNA or DNA. Smallpox virus contains DNA. Influenza virus contains RNA.

- Viruses can pass through **bacterial proof filters**. The size ranges between 90Å to 115Å.

Morphology and Structure of bacteriophages

Bacteriophage consists of head (1000Å x 650Å), tail (1000Å x 250Å), base plate and tail spikes (1300Å x 20Å). Head consists of protein capsid in which the nucleic acid is enclosed. Tail consists of protein sheath (180Å) and infection core (70Å). Tail pins are six in number and connected to the base plate. Total volume of phages is 1/1000 of a host cell. Generally, it consists of hexagonal head and filamentous tail.

The nucleic acid occurs as a continuous loop or filament composed of a single or double stranded molecule. The viral particles, termed virions may be helical or spherical or combination of both. Individual protein present in the capsid is called capsomer. Viruses enclosed with a unit membrane envelope are termed enveloped viruses. Those lacks an envelope are called naked viruses. Most animal viruses are enveloped but nearly all plant viruses and phages are naked.

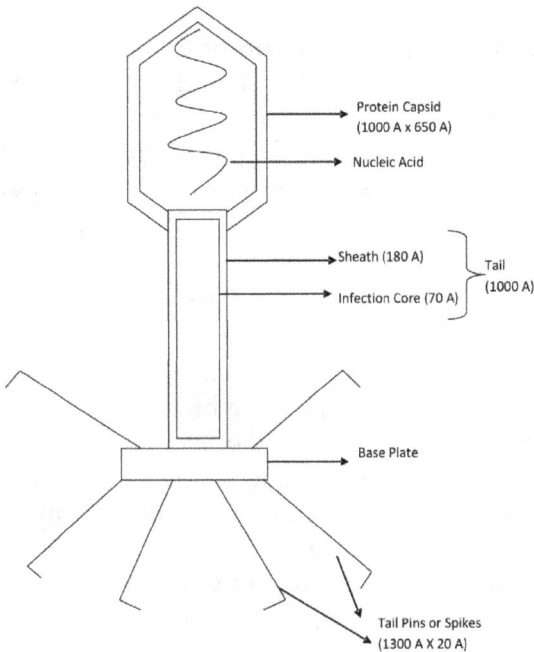

Fig. 7.1. Morphology and Structure of Bacteriophage

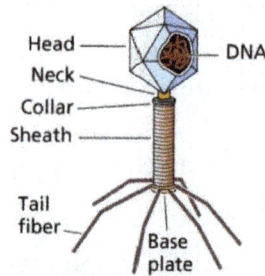

Fig. 7.2. Structure of Bacteriophage

Classification of viruses

Individual viruses are grouped on the basis of nucleic acid they contain (DNA or RNA), their size and architecture of their capsid. Other characteristics include presence or absence of an envelope and chromosome structure (circular versus linear and single versus double stranded molecules). Virion RNA that can function as mRNA is termed plus stranded RNA, while that RNA is complementary to virion mRNA is termed as minus stranded RNA.

Plant viruses are grouped on the basis of the structure of virion; whether it contains DNA or RNA or its mode of transmission.

T-Phage

The most extensively studied group of bacteriophages belongs to the T series. They are numerated from 1 to 7 and infect non motile strain of *E. coli*. All these phages are composed of DNA and protein in approximately equal amount. Except for T_3 and T_7, all of them have tadpole shape with hexagonal heads and long tails. The tail of T_3 and T_7 are short. T even phages (T_2, T_4 and T_6) have an unusual base, 5-hydroxymethylcytosine in DNA, in place of the usual cytosine.

Plaque Assay

When a virus particle initiates an infection on a layer or lawn of bacterial cells spread out on a flat surface, a zone of lyses or growth inhibition may occur that results in a clear area in the lawn of growing host cells. This clearing is called a plaque. Each plaque is originated from one viral particle and thus number of viral particles can be counted in any given sample. The various stages in viral multiplication can be shown by one step growth curve. This curve shows a single round of viral multiplication in a population of cells. Following adsorption, the infectivity of the virus particles disappears, a phenomenon called eclipse. This is due to the uncoating of the virus particles. During the latent period, replication of viral nucleic acid and protein occurs. The maturation period follows, the viral nucleic acid and protein are assembled together to form mature virus particles. Finally, release occurs, the timing of one step growth cycle varies with virus and host. For bacteriophages, it may be completed in 30-60 minutes, for animal viruses it may take 12-24 hours.

Types of bacteriophages

From infectivity point of view and host relationship, there are two types of bacterial viruses:

- Lytic or Virulent phage
- Temperate or Avirulent phage

Lytic or Virulent phage

When lytic phage infect host cell, host respond by producing large number of new virions, and at the end of incubation period the host cell bursts releasing new virions. The phase under going lytic cycle is known as lytic phage or virulent phage. Lytic development can be further divided into two general parts.

- **Early infection**-the period from entry of the DNA to the start of replication.
- **Late infection**- period from start of replication to the final step of lysis of bacterial cell to release progeny phage particles.

In early phase, production of the enzymes involved in the replication of DNA (enzymes concerned with DNA synthesis, recombination and sometimes, modification) occur. A pool of phage genomes is accumulated. In late phase, the protein components of phage are synthesized. Structural protein and sometimes, assembly proteins are synthesized.

There are five stages in lytic or virulent phage

a. Adsorption

When phages come in contact with bacterium, they get attached with the host/ bacterial cell with the help of tail pins. The base plate and spikes also help to fix the phage to the bacteria.

b. Penetration

Once the phage is attached, the sheath acts as micro-syringe. The contraction of the sheath results in injection of nucleic acid into the bacterial cell. The protein coat is left outside the bacterial cell.

c. Replication

After penetration of the nucleic acid, the synthesis of the host components and cell division is stopped. Synthesis of phage specific enzymes begin, which result in the formation of phage nucleic acid and protein particles with the help of host machinery.

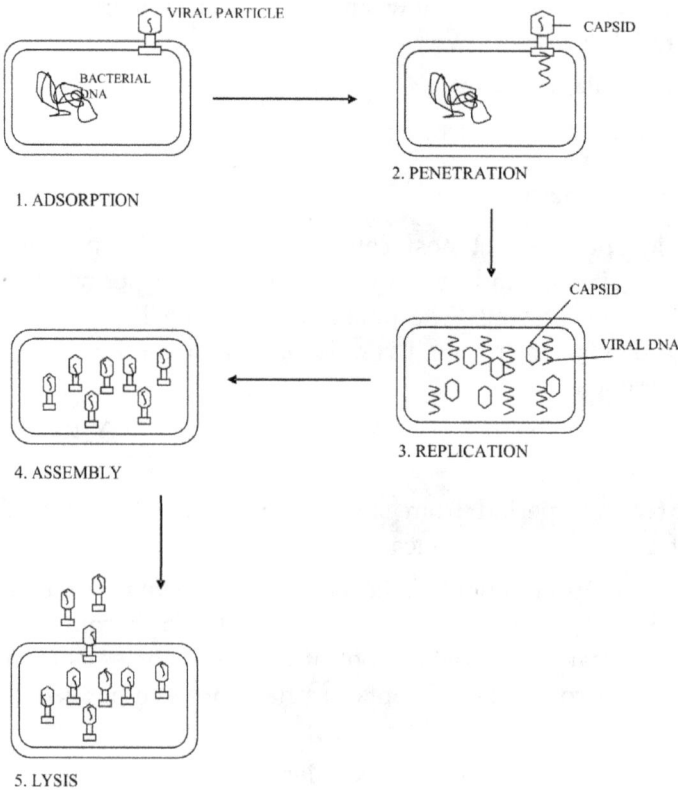

Fig. 7.3. Lytic Cycle of Bacteriophage

d. Assembly and maturation

After the phage proteins and nucleic acids are synthesized; they are assembled to form complete phage particles.

e. Lysis

When assembly is complete, lyszyme is produced. Due to the production of lysozyme, bacterial cell wall is broken down and phage particles are liberated.

Temperate phage

In it, a state of lysogeny exists and infection may not be apparent. The viral DNA is not produced by the host but transmitted genetically from generation to generation. These temperate phages may become virulent spontaneously at some subsequent generation.

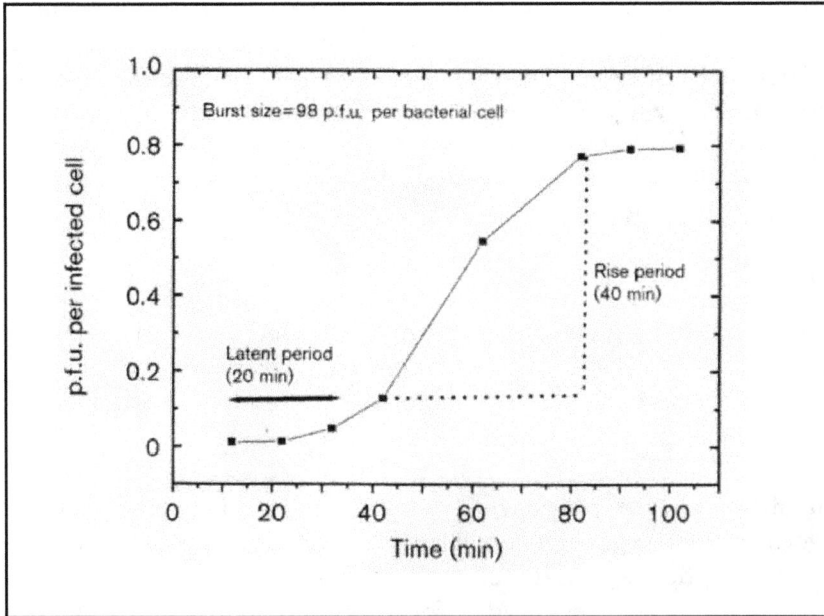

Fig. 7.4. One step growth curve

Lysogeny and induction of the temperate phage – Lambda phage

In this type of phage the adsorption and penetration stages are similar to that of lytic phage. After the entry of nucleic acid into the host cell; the viral nucleic acid associates with bacterial DNA and continue to multiply along with the host. This relationship between host and phage is known as "lysogeny". Phage which undergoes such a relationship is known as temperate phage or avirulent phage e.g. Lambda (ë) phage of *E. coli*. The bacterial strain containing the temperate phage is known as lysogenic strain or lysogen. When the viral genome is present in the bacterium in a latent form is known as prophage. The ë prophage integrates in the host chromosomes between the two sets of bacterial genes; the *gal* gene which encodes enzyme that degrade the sugar galactose ; and *bio* gene which encode biotin synthesis. Integration of the ë chromosome requires the action of a protein termed integrase encoded by the ë gene.

Lambda DNA

When λ DNA introduced into a new host cell, the lytic and lysogenic pathways start off the same way. Both require expression of immediate early and delayed early genes. But they diverge, lytic development follows if the late genes are expressed, lysogeny occurs if the synthesis of repressor gene is established. The λ has only two immediate early gene and delayed early gene; comprise of two replication genes + 7 recombination genes (two are necessary to integrate λ DNA into bacterial chromosomes for lysogeny). The CII/CIII pair of regulators are needed to start up synthesis of lysogenic repressor. The Q regulator is an antitermination factor that allows the host RNA polymerase to proceed into the late genes. So delayed early genes serve two masters: some needed for the phage to enter lysogeny, the others ensure that the lytic cycle follows its proper order.

Induction

Plaques are the region formed by phages, which are clear and circular in a turbid layer of cells (lawn). UV or chemicals damage DNA and the rec A protein is synthesized which binds to single stranded DNA(ssDNA). The rec A protein is protease which cleaves λ repressor, thus allowing viral genes to be expressed. The product of the two genes, *int* and *xis* act together to excise the prophage from the host chromosome. The phage undergoes normal vegetative replication.Usually few genes are expressed by the transcription apparatus of the host cell. These are known as early genes. One of these genes codes for protein that is necessary for the transcription of next class of genes. The second class of genes-delayed early or middle group – expression starts in the early period. As soon as the regulatory protein is synthesized, expression of host gene is reduced. When the replication of phage begins as a result of early gene expression; it also needs a regulatory protein for middle genes. Mutation in any one of the essential gene prevents successful completion of lytic cycle.

Life cycle of Plant and animal viruses

Plant and animals viruses cannot introduce their nucleic acids themselves into the host cells. Therefore, they need a vector to transfer them. Insects like mosquitoes, whitefly etc. serve as vectors in man and plants; respectively. When a vector injects virus into the host (man/animal/plant) cell, the virus undergoes uncoating inside the host cell and follow replication, assembly, maturation as it occur in the bacteriophages.

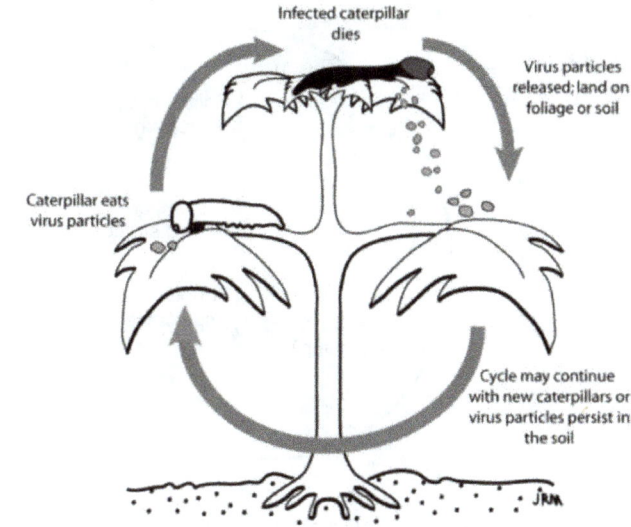

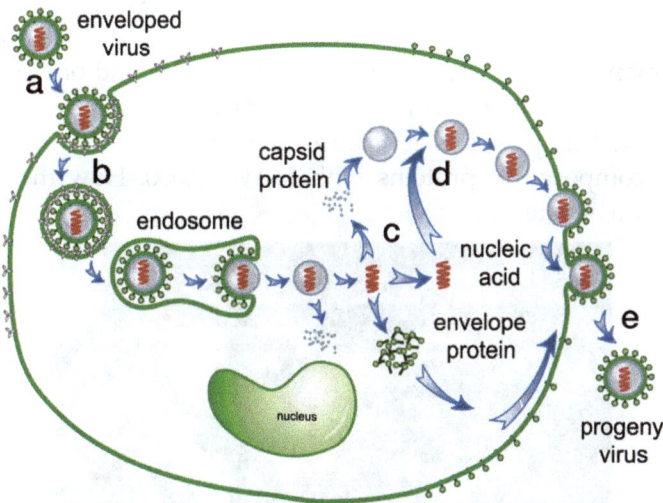

Fig. 7.5. Life cycle of Plant and Animal Viruses

Multiplication of animal viruses

Viroids

Viroids were first discovered in 1967. They are merely infectious single stranded RNA unassociated with any virion structure and cause disease in plants; exist intra-cellularly as well as extra-cellularly as circular ssRNA molecule containing about 400 nucleotides. Probably host cell RNA polymerase replicate the viroid chromosome.

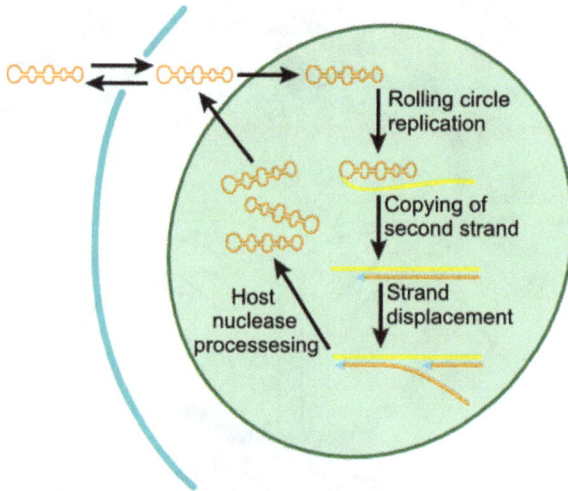

Fig. 7.6. Viroid replication

Prions

Two human diseases, Kuru and Jakob-Cerutzfeldt and one sheep disease Scrapies are caused by macromolecules, smaller than any known virus. They are termed as prions. They cause fatal degeneration of brain tissue. It is largely composed of proteins with M.W. 30,000. How they reproduce is, of course, a mystery.

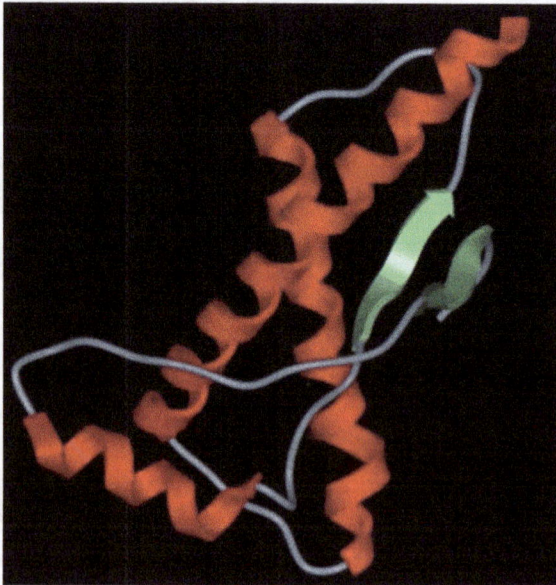

Fig. 7.7. A molecular model of a prion

Chapter 8

Classification

The place of microorganisms in the living world:

Haeckel's kingdom Protista

German zoologist, E. H. Haeckel suggested a third kingdom to include microorganisms which are typically neither plants nor animals. He proposed a new kingdom, called Protista, which comprised of only the unicellular organisms. It includes bacteria, algae, fungi and protozoa- but not viruses because they are not cellular organisms

Procaryotic and Eucaryotic cells

Microorganisms can be further divided into two categories, prokaryotes and eukaryotes. The blue green algae and bacteria are prokaryotic organisms. The eukaryotic microorganisms include the protozoa, fungi and algae. The terms lower protists and higher protists are sometimes used for prokaryotic and eukaryotic microbes, respectively.

Whittaker's five kingdom concept

Another system of classification, the five kingdom system, has been proposed by Whittaker (1969). This system is based on three levels of cellular organization and in relation to three principal modes of nutrition: photosynthesis, absorption and ingestion. The prokaryotes are included in in the kingdom Monera; they lack ingestive mode of nutrition. Unicellular eukaryotic microorganisms are placed in the kingdom Protista. The kingdom plantae included multicellular green plants and higher algae. Animalia included multicellular animals and fungi (multinucleate higher fungi).

Microorganisms are found in three of the five kingdoms: kingdom Monera (bacteria and blue green algae), kingdom protista (microalgae and protozoa) and kingdom fungi (yeasts and molds).

The Classification and Phylogeny of Bacteria

A bacterial species is a group of strains that show a high degree of overall phenotypic similarity and that differ from related strain groups with respect to many independent characteristics.

Naming of Species

According to a convention known as Binomial System of nomenclature, every biological species bears a Latinized name that consists of two words. The first word indicates the taxonomic group of immediately higher order, or genus (plural, genera) to which the species belongs and the second word identifies it as a particular species of that genus. The first letter of the generic (but not of the specific) name is capitalized, and the whole phrase is italicized: *Escherichia* (generic name) *coli* (specific name).

Numerical Taxonomy

MICHEL ADANSON
(1727–1806, French)

The taxonomic arrangement is based upon quantification of the similarities and differences among organisms. This was first suggested by Michael Adanson, a contemporary of Linnaeus, and is known as Adansonian (or numerical) taxonomy. In it, each phenotypic character is given explicit weighting, it should be possible to express numerically the taxonomic distances between organisms in terms of the number of characters they share, relative to the total number of characters examined. The determination of similarity co-efficient and matching coefficient for two bacterial strains can be calculated/determined.

No. of characters positive in both strains: **a**
No. of characters positive in strain 1 and negative in strain 2: **b**
No. of characters negative in strain 1 and positive in strain 2: **c**
No. of characters negative in both strains: **d**

$$\text{Similarity coefficient (sj)} = \frac{a}{a+b+c}$$

$$\text{Matching coefficient (Ss)} = \frac{a+d}{a+b+c+d}$$

The data can then be transposed into a dendrogram

Base composition of DNA

The melting temperature of DNA is the temperature at which it becomes denatured by breakage of hydrogen bonds and is directly related to G+C content. The G+C content varies in different species of a genera; hence it is one of the criteria for classification of organisms.

NUTRITIONAL CLASSIFICATION

Nutritional classification is based on two parameters, the nature of the energy source and nature of the principal carbon source. With respect to energy source, there is a basic dichotomy between organisms. Microbes which are able to use light as an energy source, are termed phototrophs, and organisms that are dependent on a chemical energy source, termed chemotrophs. Organisms which are able to use CO_2 as principal carbon source are termed as autotrophs and that are dependent on organic carbon source are termed as heterotrophs.

On the basis of above mentioned criteria, four major nutritional categories have been identified.

1. Photo-autotrophs

Those organisms which use light as energy source and CO_2 as the carbon source are called photo-autotrophs e.g. many photosynthetic bacteria.

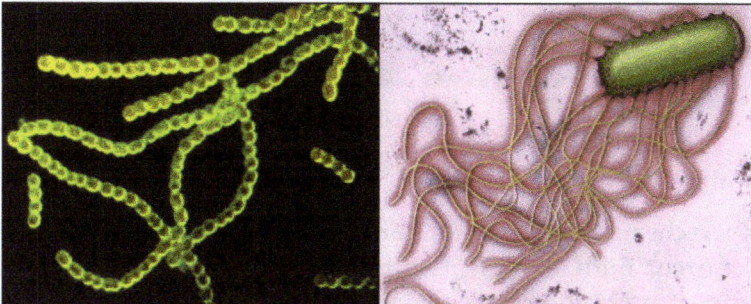

Fig. 8.1. Photosynthetic Bacteria

2. Photo-heterotrophs

They use light as the energy source and an organic compound as the principal carbon source. This category includes certain purple and green bacteria.

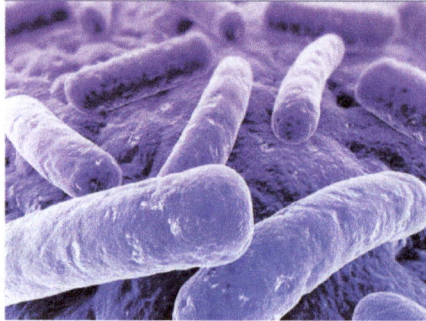

Fig. 8.2. Purple Bacteria

3. Chemo-autotrophs

The organisms which use chemical energy source and CO_2 as the principal carbon source. Energy is obtained by oxidation of reduced inorganic compounds, such as NH_4^+, NO^-_2, H_2, H_2S, S, $S_2O_3^{-2}$, CO or ferrous ion. As these bacteria have the ability to grow in strictly mineral media, they are also termed as chemolithotrophs e.g. nitrifying bacteria like *Nitrosonomas, Nitrobactor,Thiobacillus* etc.

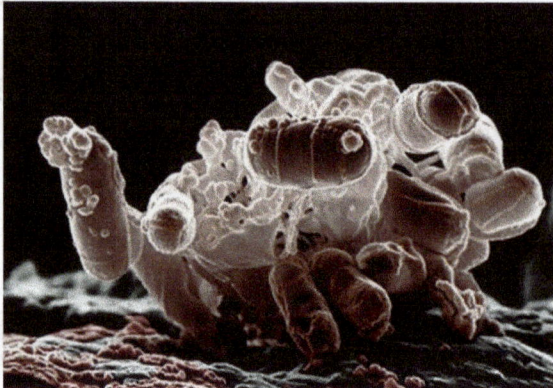

Fig. 8.3. Nitrosomonas

4. Chemoheterotrophs

They use chemical energy source and an organic substance as the principal carbon source. Both carbon as well as energy can be derived from metabolism of a single organic compound. Most of the bacteria belong to this category and bacteria take up all nutrients in dissolved form, they are also called osmotrophs *e.g. E.coli, Salmonella, Pseudomonas* etc.

The qualifications of obligate aerobes (need O_2 for respiration and cannot survive without it) and facultative aerobes (aerobic but also grow under anaerobic environment) are often used to indicate the absence (or presence) of nutritional versatility. On the basis of requirement for growth factors, the bacteria can also be classified as prototrophs or auxotrophs. A prototrophs can derive all carbon requirements from the principal carbon source. An auxotroph requires, in addition to the principal carbon source, one or more organic nutrients (growth factors) for its growth.

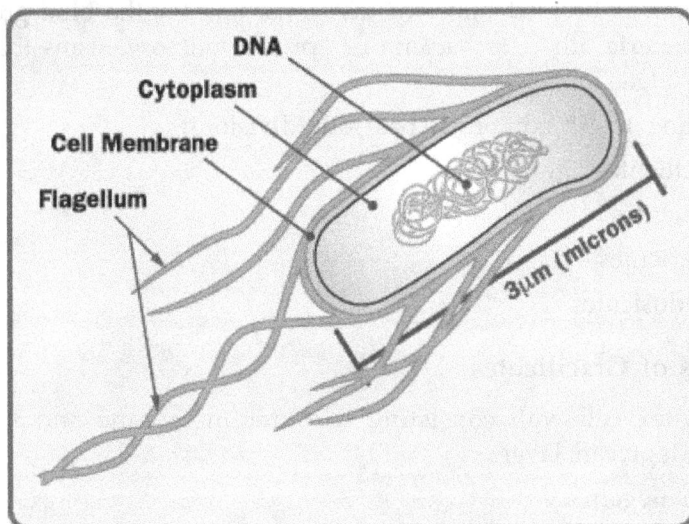

Fig. 8.4. E.coli Bacterium

Kingdom: Prokaryote

Prokaryotes (or *monera*) are the simplest living things; *bacteria* and *archaea*. They generally do not have a *cell nucleus*, nor cell *organelles*, however a small number of exceptions have been found. The name comes from *Greek pro-* (meaning *before*) and *karion*, meaning *nut* or *kernel*. Prokaryotes are cells that do not have a cell *nucleus*, and lack other things *eukaryotes* (cells with a true nucleus) have. Prokaryotes do not have *membranes* inside the cell. This means that there are no *vacuoles, Golgi apparatus, endoplasmic reticula* or other organelles inside the cell.

Prokaryotes are unicellular. They are either *bacteria* or *archaea*. The Archaea include simple *organisms*, that were first discovered in extreme environments. Most of them can survive at very high or very low temperatures. Some of them can also survive in highly *salty, acidic* or *alkaline* water. Some have been found in *geysers, black smokers* or *oil wells*.

Cells with a nucleus are called *Eukaryotes*. Eukaryote cells include *organelles* which were once free-living prokaryotes. These are like cells but they do not have any organelles, they actually infect in the cells, a virus infects a cell by attaching to it and injecting viral nucleic acid. Eventually the nucleic acid directs the cell to make more viral nucleic acid and protein coats.

Bergey's Manual of Determinative Bacteriology

The 8th edition of Bergey's Manual of Determinative Bacteriology recognized the kingdom Monera of Whittaker but has called it Prokaryotae. This kingdom is divided into two divisions, one for the blue green algae or **Cyanobacteria** other for bacteria or "prokaryotic organisms that are not blue green algae".

According to 9th edition, it has four divisions:

1. Gracilicutes
2. Firmicutes
3. Tenericutes
4. Mendosicutes

Characters of Gracilicutes

- Complex cell wall consisting of outer membrane and inner thin peptidoglycan layer.
- Gram negative.
- Cell shape- sphere, rods and filaments.
- Sheath or capsule may be present.
- Reproduction by binary fission, however budding and multiple fission rarely occur.
- No endospore formation.
- Motile and non-motile.
- Aerobes, anaerobes and facultative anaerobes.
- Phototrophs and nonphototrophs.
- Fruiting body and myxospore formers.
- Obligate intacellular parasites.

Important groups of Gracilicutes

- Aerobic Gram rods and cocci.
- Facultative anaerobic Gram negative rods.

- Spirochetes.
- Gram negative oxy-photobacteria.
- Chemoautotrophic bacteria.

GROUP-I

It has four families

1. Family Pseudomonaceae: Members of this family are straight or curved rods, motile by polar flagella, present in soil and water. Aerobic, some can grow anaerobically using nitrate as electron acceptor. Many species are pathogenic for human and animals, some cause spoilage of meat and other foods. Some species can use above 100 compounds as carbon source. *P. aeruginosa* produce a blue pigment, pycoyanin. It is mainly soil organism but it is opportunistic organism also as it can be isolated from wounds, burns etc.

2. Family Azobacteriaceae: Large cells, rod shape to oval, motile by polar or peritrichous flagella, heterotrophic, cyst former; present in soil and plant rhizosphere, can fix nitrogen under aerobic conditions.It includes four genra, *Azotobacter, Azomonas, Beijerinckia* and *Derxia*. *Azotobacter* forms cyst. High rate of respiration use oxygen rapidly at cell surface and maintain anaerobic conditions inside the cell to protect oxygen sensitive nitrogenase.

3. Family Rhizobiaceae: Rod shaped cells, motile with one polar or subpolar flagellum or 2-6 peritrichous flagella; form root nodules, stem nodules or tumors e.g. *Rhizobium, Bradyrhizobium* and *Agrobacterium*. Bacteria attach to root hairs, penetrate the roots, proliferation of root cells occurs and nodule is formed. Within the nodule, bacteria are present as bacteriods. Leghaemoglobin-a pigment is also present in root nodule and help to protect nitrogenase enzyme from oxygen damage. The bacteria show host specificity. According to recent classification, it includes

- *Rhizobium*- fast growing species
- *Bradyrhizobium*- slow growing species
- *Azorhizobium*- form stem nodules
- *Agrobacterium*- It does not fix nitrogen, form tumors on roots of dicot plants. It invades roots and stems of many di-cotyledons and gymnosperms. Tumor formation is associated with that of Ti plasmid. Strains without Ti plasmid are not tumerogenic. During infection, a Ti plasmid is transferred to plant cell; a part of it is integrated into nuclear material of transformed plant cells. The transformed cell produces opines. This creates a unique ecological niche for agrobacteria as they can use opine or C and N source for their growth.

The plasmid of *Agrobacterium tumefaciens* is used as biotecnilogical tool for gene transfer

4. Family Acetobacteriaceae: Ellipsoidal to rod shaped cells, oxidize ethanol to acetic acid e.g. *Acetobacter*. They have peritrichous flagella and are used to make vinegar. *Gluconobacter* has polar flagella, used in manufacture of chemicals, dihydroxyacetone, sorbose etc. Both the genera are present in sugar and alcohol enriched environments like flowers, fruits, vinegar, souring fruit juices etc.

GROUP-II

It include one family

1. Family Enterobacteriaceae: It is the largest well defined group of Gram negative eubacteria, small rods which may be motile by peritrichous flagella or non motile. Distinguishing property of this family is facultative anaerobic in nature. They can use wide range of organic compounds as substrate for respiration. Fermentative energy is produced by fermentation of CH_2O. *Escherichia coli* is the normal flora of intestine of mammals; may be pathogenic and cause urinary and intestinal infections. *Salmonella* and *Schigella* are closely related to *E. coli* and all the three genera are commonly called coliform bacteria, responsible for causing bacterial dysentery, typhoid fever, food poisoning and also cause intestinal infections. *Enterobacter*, *Serratia* and *Proteus* are closely related to coliform bacteria but in ecology, they are different as they are habitant of soil and water. *Erwinia* is a plant pathogen. Together with coliform, they are called enteric group of bacteria. *Yersinia* is also included in this group. *Yersinia pestis* cause plague. Some polar flagellated bacteria like *Vibrio, Aeromonas, Photobacterium* are also included in this group. G+C content varies from 37-63 per cent. During fermentation they produce lactic acid, acetic acid, succinic acid, formic acid, carbon dioxide, water and ethanol; but ratio of end products varies. Gas formation as a result of sugar fermentation is important character in this enteric group as it distinguishes gas former *E. coli* from pathogenic *Schigella* and *Salmonella typhi* which do not produce gas.

Coliform bacteria are important in sanitary analysis. Enteric diseases caused by coliform are transmitted by faecal contamination of water and foods. Cholera and typhoid fever spreads through contaminated water. Coliform bacteria are used as indicator bacteria.

GROUP-III

Family Spirochaetes

- Slender, helical coiled forms.
- May be aerobic, anaerobic and facultative anaerobic.

- Free living or parasitic.

- Some species are pathogenic.

GROUP-IV-Gram negative oxyphotobacteria, or oxygenic phototrophic bacteria or cyanobacteria

Gram negative phototrophic bacteria have 35-71 per cent G+C content, oxygenic photosynthesis and fix nitrogen. Oxygen sensitive nitrogenase is present. Nitrogen assimilation occurs in specialized cells, called heterocyst. Unicellular to filamentous phoyosynthetic apparatus is located in thylacoids. Heterocyst has chlorophyll a, Phycobilin protein, PS II and ribulose biphosphate, but carboxylase – key enzyme of calvin cycle is absent. Cells in filaments are connected by minute channels called microplasmadesmata for exchange of metabolites. Examples are *Nostoc, Anabaena,Oscillatoria, Pleurocapsa* and *Spirulina*. They also form specialized resting cells akinetes which germinate to form new filaments under favorable conditions. Reproduction also occurs by harmogonia. *Anabaena* form symbiotic association with *Azolla*, an important fern that grows on the surface of still water and fix nitrogen in rice fields. Mucilage containing cavities are present on the leaves of Azolla which are initially open; but later closed and harbor filaments of *Anabaena*. *Azolla* is used to fix nitrogen in rice field.

Group-V- Gram negative anoxyphotobacteria

It includes Green sulfur and Green non sulfur bacteria.

Differences between Green sulfur and Green non sulfur bacteria

Green sulfur bacteria	Green non sulfur bacteria
1. Straight or curved rods.	1. Filamentous gliding bacteria.
2. Sulfur deposition extracellular	2. Thermophile able to grow 45-70^{0}C.
3. Strictly anaerobic phototrophic.	3. Photoheterotrophs and facultative phototrophs.
4. Use hydrogen sulfide and other reduced compound as electron donars. They synthesize β-hydroxybutyrate. e.g. *Chlorobium*.	4. Grow in hot springs that have low content of organic matter. Synthesize-hydroxcybutyrate e.g. *Chloroflexus*.

Group-VI- Chemolithotrophic bacteria

Aerobic, Gram negative, grow strictly on mineral medium, drive carbon from carbon dioxide. ATP and reducing power are derived from respiration of inorganic compounds. These were first of all discovered by Winogradsky.

Two important characteristics of this group are

a. High specificity with respect to inorganic energy source.

b. Inability to use organic compound as carbon and energy source. Utilizable substrates are hydrogen sulfide and reduced compounds of sulfur, ammonia, nitrate, molecular hydrogen CO and ferrous ions.

Various subgroups are as follows

- **Nitrifying bacteria**: *Nitrococcus* and *Nitrosomonas*.
- **Sulfur oxidizers**: *Thiobacillus*.
- **Iron bacteria**: *Galleonella, Thiobacillus ferroxidans*
- **Hydrogen bacteria**: *Hydrogenomonas*
- **Carboxybacteria**: *Carboxy mmas*

II Divison: Firmicutes

Characters of this division are:

- Cell wall with thick peptidoglycan layer.
- Generally Gram positive.
- Cell shape sphere, rods and filaments.
- Reproduction by binary fission.
- Endospores or spores on hyphae are formed by some species.
- Aerobes, anaerobes and facultative anaerobes.
- Chemo-heterotrohic and non-photosynthetic.

Important groups of this division are:

- Endospore forming bacteria.
- Actinomycetes.
- Lactic acid bacteria

It has ten groups:

GROUP-I

- Gram positive cocci, spherical cells.
- Aerobic, anaerobic and facultative anaerobic.

Some of the important genera of this group are as follows:

- *Micrococcus*- common inhabitant of soil, fresh water and skin of man and other mammals.
- *Staphylococcus*- Grow in presence of 15 per cent NaCl, mainly associated with skin or mucous membrane of warm blooded animals.

- Other important genera are *Streptococcus, Leucon stoc, Pediococcus,* and *Sarcina.*

GROUP-II - Endospore forming rods and cocci

Spore formation is a complex process of differentiation that is initiated as the population passes out of exponential phase and approaches stationary phase. Some of the important genera of this group are as follows

- *Bacillus-* rods, aerobic, some are facultative anaerobic.
- *Clostridium, Desulfomaculum-* rods, strictly anaerobic.
- *Sporolactobacillus-* rods and aero-tolerant.
- *Thermoactinomycetes-* mycelia, aerobes.

GROUP-III- Non-endospore forming Gram positive rods

- Rod shaped.
- Examples are *Lactobacillus* and *Listeria* used in dairy products.

GROUP-IV Non-endospore forming bacteria of irregular shape

This heterogeneous group has common features as given below

- Straight or curved rods do not have uniform shape, show swellings, club shaped and other deviations. Aerobic or facultative aerobic, exhibit respiratory metabolism, also show fermentative type of metabolism.
- *Corynebacterium* is rod shaped cells, show club shaped swellings, club shaped and accumulate intracellular volutine granules, which gives reddish purple color with methylene blue dye. Cell wall contains mycolic acid (corynemycolic acid). *Corynebacterium diptheriae* causes diptheria in human beings.
- *Brevibacterium* -rod-coccus cycle, orange colonies and salt tolerant, present on cheese surface where it produces proteolytic enzymes which helps in cheese ripening process.

The other examples are *Arthrobacter, Cellulomonas* and *Microbacterium*

GROUP-V - Aerobic Nonfilamentous or Filamentous Rods

- *Microbacterium-* small slender, irregular rods do not show rod coccus cycle, but rods are shorter in stationary phase, present in milk and dairy products.
- *Propionibacterium* has pleomorphic morphology, produces propionic and acetic acid and are present in dairy products on human skin and in the intestine.

- *Eubacterium* has pleomorphic morphology, produces butyric and formic acid and are present in intestinal tract and on skin; but it is nonpathogenic.
- *Actinomyces* have pleomorphic morphology, it cause actinomycosis in human beings.

GROUP-VI -Mycobacteria

- Most of the members are pathogenic.
- Straight, curved or straight rods.
- Presence of mycolic acid and acid fast cell wall.

  e.g. *Mycobacterium tuberculosis* and *Mycobacterium leprae.*

GROUP-VII - Nocardiforms

- Aerobic bacteria, surface mycelium are formed by fragmentation of rods or coccus forms.
- Some also contain aerial mycelium which develop conidiospores.

  e.g., *Nocardia.*

GROUP-VIII- Actinomycetes

- Dividing in more than one plane.
- They form spores; form a tissue like mass of cells derived by divisons in different planes e.g. *Dermatophilus, Geodermatophilus* and *Frankia. Frankia* is symbiotic and filamentous mycelial bacteria, forms root nodules in nonleguminous dicotyledonous plants and fixes atmospheric nitrogen.

GROUP-IX -Streptomycetes and their allies

A well branched mycelium is formed that does not fragment readily. Reproduction is by spore or sometime by growth of mycelial fragments. They produce variety of antiboitics e.g. *Streptomyces.*

GROUP-X -Other conidiates

- Members produces both aerial and substrate mycelium except in *Micromonospora.* Sporophores are lacking or are very short. e.g. *Actinopolyspora, Thermomonospora, Thermoactinomyces* and *Micromonospora.*

Divison Tenericutes

Distinguishing characteristics are as follows

- Lack cell wall and do not synthesize peptidoglycan precursor.

- Enclosed by a unit membrane plasma membrane and stain Gram negative
- Cell shape pleomorphic, may appear as larger, irregular vesicles to filamentous forms.
- Reproduction by budding, fragmentation and /or binary fission.
- Mostly non motile, show gliding movement.
- Require complex media for growth. Inclusion of cholesterol and long chain fatty acids (chemoheterotrophic) in growth medium is necessary for their isolation.
- Growth on solid media is characterized by the formation of fried egg colonies.
- Saprophytes; and pathogens of animals, plants and human beings.
- Completely resistant to beta lactam antibiotics.

Mycoplasma: It has following important characters like absence of cell wall, obligate chemoheterotrohs with complex nutritional requirements. Mostly parasitic on plant and animals. Reproduction is by binary fission. They require sterol and cholesterol for growth. They are insensitive to penicillin e.g., *Mycoplasma*.

Divison Mendosicutes

Distinguishing characterstics are as follows:
- Cell wall that does not contain true peptidoglcan but have psuedo – peptidoglycan.
- Cell wall may include large protein molecules or variety of polysaccharides.
- Cell shape cocci, rods, filaments and those resembling mycolpasma.
- Many are motile.
- Endospores or other resting spores are not reported.
- Most species are strict anaerobes.
- Ecologically and metabolically diverse and capable of living in extreme environments.

Important group is Archaebacteria

GROUP- Archaebacteria

This group is divided into three subgroups : Methanogens, Halophiles and Thermoacidophiles.

1. Methanogens

The methanogens convert carbon-dioxide, formate and acetate to methane. Their cell wall is made up of pseudomurein and N-acetylmuramic acid and lacks D- amino acids. Various examples are *Methanobacterium* and *Methanobrevibacterium*. *Methanococcus* have flexible cell wall composed of protein with traces of glucosamine. *Methanosarcina* is made up of acidic heteropolysaccharides. *Methanospirillum* have complex cell wall of two layers inner of unknown chemical composition and outer layer is made up of protein. Protein is resistant to hydrolysis by protease.

2. Halophiles

Present in saline environment. It is of two types. Immotile cocci-*Halococcus* and polar flagellated rods-*Halobacterium*.

- Minimum NaCl concentration is 2.0 to 2.5 M

  Optimum NaCl concentration is 4.0 to 5.0 M

- Magnesium requirement is also very high.

- Characteristic feature is presence of red color carotenoids which are incorporated into the cell membrane which protect cell from photochemical damage by high light intensities (characteristics of natural environment).

- Aerobic heterotrophs with complex nutritional requirements.

- Cell wall of *Halobacterium* are large acidic glycol-proteins. In addition to this, they also contain non-glycosylated protein and glycolipids. Cell wall of *Halococcus* is made up of heteropolysaccharide.

3. Thermoacidophiles

It grows at high temperature and low pH. They are of three types namely *Sulfolobus*, *Thermoplasma* and *Thermoproteus*.

Sulfolobus

- Grow at high temperatures in the range of 55-85°C.
- Average pH range is 1.0 to 5.9.
- Cell wall is made up of lipoprotein and CH_2O(carbohydrate).

Thermoplasma

- Facultative anaerobes that use small number of mono and disaccharide as a source of carbon and energy.
- All strains have absolute requirement for an unusual growth factor provided from yeast extract.

- They are found only in refuse piles from coal mines which contain residual coal and substantial amount of iron pyrite.
- They lack cell wall but cell membrane contain large amount of LPS and glycoprotein.

Thermoproteus

- Use elemental S as terminal electron acceptor.
- Strict anaerobes.
- Cell wall consists of glycoprotein e.g. *Thermoproteus* and *Desulfurococcus.*

Chapter 9

Biogeochemical Cycles

The term soil refers to the outer, loose material of the earth's surface, a layer distinctly different from the underlying bedrock. The soil is composed of five major components, mineral matter, water, air organic matter and living organisms. Air and water together account for approximately half the soil's volume, which is known as pore space. The proportion of air and water

fluctuates depending upon the circumstances. The mineral fraction contributes slightly less than half the volume. Organic matter usually contributes some three to six per cent of the total. The living portion of the soil body makes up slightly less than one per cent of the total volume. The organic constituents of plants are commonly divided into six broad categories:

(a) The most abundant chemical constituent, cellulose, varying in quantity from 15 to 60 per cent of the dry weight.

(b) **Hemicelluloses**, making up 10 to 30 per cent of the weight.

(c) **Lignin**, which usually makes up 5 to 30 per cent of the plant.

(d) **Water soluble fraction**, like simple sugars, amino acids, and aliphatic aids, these contributing 5 to 30 per cent of the tissue weight.

(e) **Alcohol and other soluble constituents**, fraction containing fats, oils, waxes, resins and a number of pigments and proteins.

(f) The mineral constituents vary from 1 to 13 per cent of the total tissue.

Carbon Cycle

Organic matter decomposition serves two functions for the microflora; providing energy for growth and supplying carbon for the formation of new cell material. The cells of most microorganisms commonly contain approximately 50 per cent carbon. The source of the element is the substrate being utilized. The process of converting substrate to protoplasmic carbon is known as assimilation. Under aerobic conditions, frequently from 20 to 40 per cent of the substrate carbon is assimilated; the remainder is released as CO_2 or accumulates as waste products. The fungal flora generally releases less CO_2 for each unit of carbon transformed aerobically than the other microbial groups because the fungi are more efficient in their metabolism. During decomposition by fungi, some 30 to 40 per cent of the carbon metabolized is used to form new mycelium. Populations of many aerobic bacteria, less efficient organisms, assimilate 5 to 10 per cent while anaerobic bacteria incorporate only about 2 to 5 per cent of the substrate carbon into new cells.

As carbon is assimilated for generation of new protoplasm, there is concomitant uptake of nitrogen, phosphorus, potassium and sulfur. It leads to immobilization of these elements. Hence, immobilization is a mechanism by which microorganisms reduce the quantity of plant available nutrients in soil. The rate of decomposition of plant materials depends upon the nitrogen content of the tissues, protein rich substrates being metabolized most readily. As crop plants generally contain about the same amount of carbon, usually about 40 per cent of the dry weight, their nitrogen contents can be compared by use of the C:N ratio. Thus, low nitrogen content or a wide C:N ratio is associated with slow decay.

During mineralization, the C:N ratio tends to decrease with time. This result from the gaseous loss of carbon, therefore, the per centage of nitrogen in the residual substance continuously rises as decomposition progresses. The C:N ratio for humus being roughly 10:1. Microorganisms play important role in soil by various transformations of organic and inorganic forms of major and minor elements essential for plant growth. Soil is a storehouse of microbes, for example one gram of soil contains up to 10^8 bacteria, 10^5 fungi and 10^4 actinomycetes. Some of the important transformations brought about by microbes include elements like Nitrogen, Carbon, Sulfur and Phosphorus.

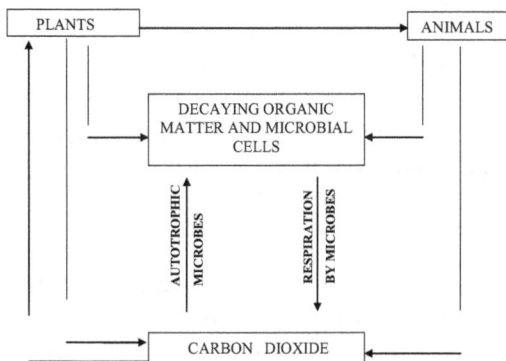

Fig. 9.1. Graphic Sketch of Carbon Cycle

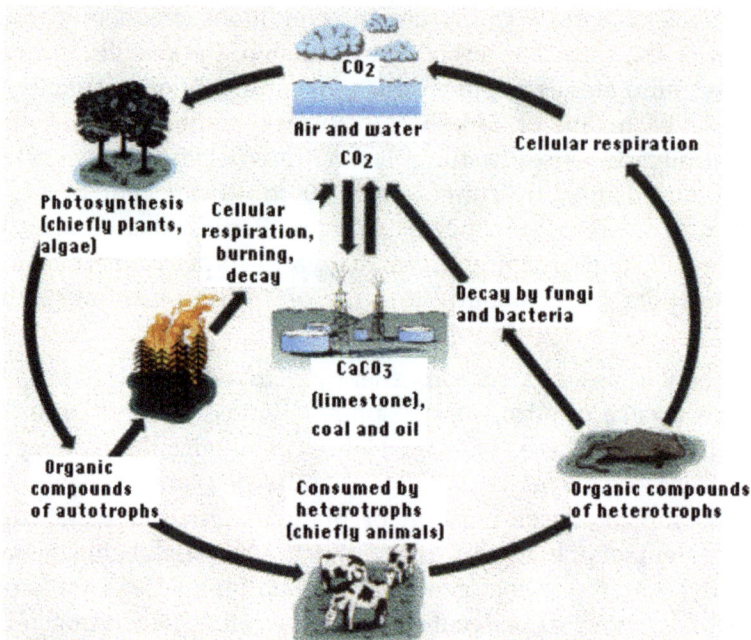

Fig. 9.2. Detailed diagram of Carbon Cycle

Carbon cycle is unusual among nutrient cycles in that it is not necessarily have to involve decomposers. Plant cells and microbial cells contain large quantities of carbon (about 40-50 per cent). It exists as carbon dioxide in atmosphere having concentration 0.03 per cent by volume. Carbon dioxide is excreted by all organisms in respiration. Carbon sink is fossil fuels. Carbon in fossil fuels can be returned to the carbon cycle by burning. Carbon dioxide is also dissolved in water in the form of bicarbonate ions. Carbon is constantly fixed into organic form by photosynthetic organisms and plants. Vegetation of the earth surface fixes 1.3×10^4 kg carbon dioxide per year. The total mass of carbon currently in atmosphere is about 700,000 million tones. Rate of decomposition of plant materials depend upon the nitrogen content of the tissues. The wider the C:N ratio, slower is the decomposition. When C:N ratio comes down to 10:1, then the organic carbon is completely mineralized and is no longer necessary for microbial growth. Carbon in plant material is mainly present in the form of polysaccharides which include cellulose, hemi-cellulose, lignin and starch.

Cellulose

Cellulose is a carbohydrate composed of glucose units bound together in a long, linear chain by α-linkage at carbon atoms 1 and 4 of the sugar molecule. It is localized in the cell wall as submicroscopic rod shape units known as micelles. The micelles are further arranged into a larger structure, the micro-fibrils, which may contain 10-20 micelles. A number of polysaccharides are associated with the cellulose of the plant cell wall. These include xylans, mannans and polyuronides. These polysaccharides are known as cellulosans.

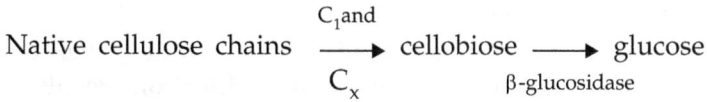

$$\text{Native cellulose chains} \xrightarrow[C_x]{C_1 \text{and}} \text{cellobiose} \xrightarrow{\beta\text{-glucosidase}} \text{glucose}$$

Cellulose is broken down by cellulase enzyme complex which is composed of C_1, C_x and β-glucosidase enzymes.

Hemi-cellulose

Hemi-cellulose is water insoluble polysaccharide extracted from plants by dilute alkalis. On breakage it gives hexoses, pentoses and uronic acids. Hemi-cellulose is mainly present in the middle lamella of cell wall. They are also known as pentosans, hexosans, polyoses or noncellulolosic cell wall polysaccharides. Pentoses are mainly xylans and arabans, composed of polymers of xylose and arabinose, respectively; while hexoses are mannan and galactan composed of mannose and galactose subunits. Hemi-cellulose is degraded by hemi-cellulase enzyme complex.

Lignin

The third most abundant constituent of plant material is lignin. It is found in the secondary layers of the cell wall and to some extent in the middle lamella. It is difficult to be degraded because of its aromatic nature. It also provides strength to the plant. The basic unit of lignin is phenyl propane.

Starch

Plant starch usually contains two components – amylose and amylopectin. The former has a linear structure built up of 200 to 500 glucose molecules linked by α - 1,4 glucosidic bonding. In amylopectin, the molecule is branched and has side chains attached through α -1,6 linkages. Starch is decomposed by amylase enzyme which has two subunits - α and β-amylases. The α – amylase acts on only amylose, while β-amylase acts on both amylose as well as amylopectin.

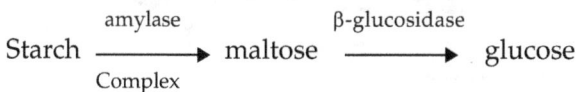

$$\text{Starch} \xrightarrow[\text{Complex}]{\text{amylase}} \text{maltose} \xrightarrow{\beta\text{-glucosidase}} \text{glucose}$$

Nitrogen Cycle

Nitrogen is one of the major elements required by the plant which is present in inert state in the environment. It is to be reduced to ammonia before it is utilized. Nitrogen reaches to soil through

(a) rain water;

(b) biological nitrogen fixation;

(c) through fertilizer incorporation.

Plants use nitrogen from soil. Plants and animal residues when buried into soil are subjected to proteolysis and ammonification, resulting in formation of ammonia. Ammonia either absorbed by plants or converted to nitrate through nitrification.

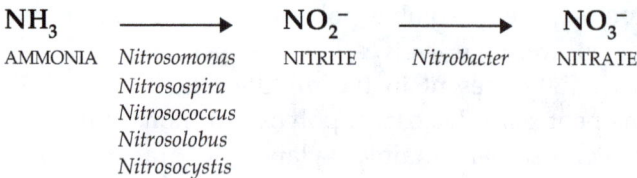

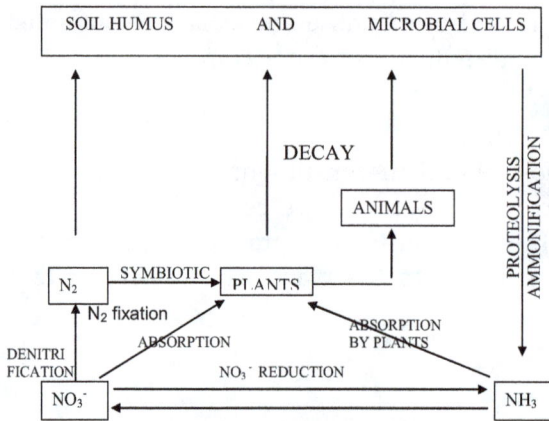

Nitrification

$$NH_3 \longrightarrow NO_2^- \longrightarrow NO_3^-$$

AMMONIA	*Nitrosomonas*	NITRITE	*Nitrobacter*	NITRATE
	Nitrosospira			
	Nitrosococcus			
	Nitrosolobus			
	Nitrosocystis			

Fig. 9.3. The Nitrogen cycle

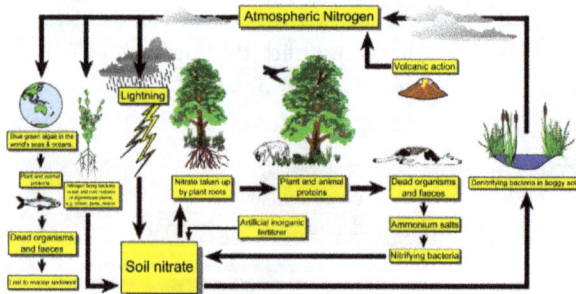

Fig. 9.4. Graphic Sketch of Nitrogen Cycle

Proteolysis

Enzymatic degradation of protein is known as proteolysis. Different microorganisms involved are *Pseudomonas, Bacillus, Clostridium, Aspergillus, Proteus* and *Penicillium.*

Ammonification

End products of proteolysis are amino acids. Microbes disseminate amino group and produce ammonia. Different microorganisms involved are *Pseudomonas, Bacillus* and *Aspergillus.*

Nitrification

Oxidation of ammonia to nitrate is called nitrification. Different microorganisms involved are *Nitrosomonas, Nitrospira, Nitrococcus, Nitrosolobus* and *Nitrobacter.*

$$NH_3 \rightarrow NH_2OH \rightarrow NHOH \rightarrow NO_2^- \rightarrow NO_3^-$$

Nitrate Reduction

Conversion of nitrate to ammonia is called nitrate reduction. Different microorganisms involved are *Azotobacter* and *Pseudomonas.*

$$NO_3^- \rightarrow NH_3$$

Denitrification

Conversion of nitrate to nitrogen is called denitrification. It occurs in anaerobic conditions. Different microorganisms involved are *Pseudomonas denitrificans, Thiobacillus denitrificans* and *Micrococcus denitrificans.*

Biological Nitrogen Fixation

In the environment, nitrogen concentration is 78 per cent by volume. Conversion of atmospheric elemental nitrogen into ammonia through a reductive process with the help of microbes is known as biological nitrogen fixation. In nature, 70 per cent of total nitrogen is fixed by biological means and rest by chemical means. It is of three types:

(a) Asymbiotic Nitrogen Fixation

In it nitrogen is fixed under free living conditions. Amount of nitrogen fixed is up to 50 kg N/ha/year. Different microorganisms involved in nitrogen fixation are *Azotobacter, Clostridium, Rhodospirillum, Anabaena* and *Nostoc.*

(b) Symbiotic Nitrogen Fixation

Bacteria which fix nitrogen in symbiosis with legume crops belong to genus *Rhizobium* which penetrates into root hair and form nodules in legumes. Efficient nodules are pink in color. Enzymatically, nitrogen is converted to ammonia with the help of enzyme nitrogenase. Rhizobia are divided into three groups

- Fast growing rhizobia
- Slow growing rhizobia or bradyrhizobia
- Stem nodule forming rhizobia or azorhizobia

(c) Associative Nitrogen Fixation

Bacteria adhere to the root surface of plant and fix nitrogen e.g. *Azospirillum.*

Nodulation process of *Rhizobium*

First of all, the legumes secrete root exudates in their vicinity to which host specific *Rhizobia* are attracted. This is followed by root hair curling and invasion of root hair by host specific *rhizobia*. Indole acetic acid (IAA) and lectins are possibly concerned in this process. Following the microbial penetration into the root hair, a hyphae like infection thread is formed. The bacteria are released into the cortical region of root system. Following the release, a period of rapid cell division takes place in the host cells. The cortical cells into the nodule region become tetraploid. Efficient nodules are pink in color due to presence of leg-haemoglobin. The nodules are rounded, lobed or club shaped depending upon the host. Infection thread branch and distribute themselves over the tetraploid cells. The root nodule results from tissue proliferation induced by the *rhizobia* via growth hormones. Once liberated from the infection thread, the *rhizobia* assume a peculiar morphology, called bacteriods. These bacteroids proliferate rapidly and are irregularly shaped.

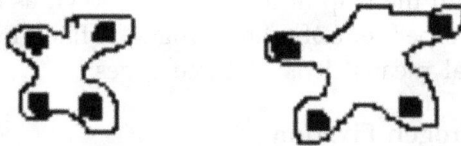

Fig. 9.5. Structure of Bacteroids

The root nodules formed by the bacteria on legumes fix atmospheric nitrogen and fulfill the nitrogen requirements of leguminous plants. Nodules formed by an efficient strain of *Rhizobium* meet the whole nitrogen requirement of the plant and there is no need to supply nitrogen by other means. The legumes excrete excess amount of organic nitrogen into soil to nourish the succeeding crop. In return to nitrogen fixation, the bacteria get protection, proper conditions for growth and photosynthates as source of energy. It may be mentioned here that neither the bacterium nor the plant can fix atmospheric nitrogen independently.

Root nodules in non-leguminous plants

Many higher plants which are not members of leguminosae, also form root nodules with the ability to fix nitrogen. In most cases, these endosymbionts are actinomycetes belonging to genus *Frankia.* The host plant of such actinomycetes includes *Casurina, Albus, Myrica* and *Dryas* etc.

Ammonia formed through nitrogen fixation is assimilated into plant system using following steps:

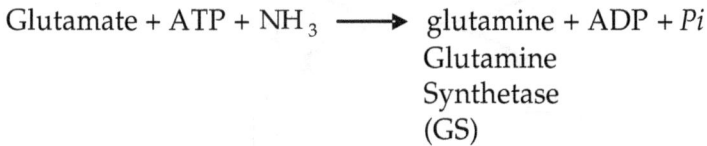

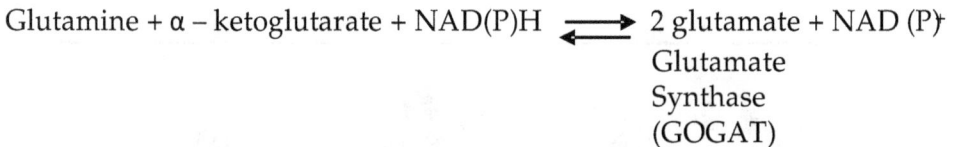

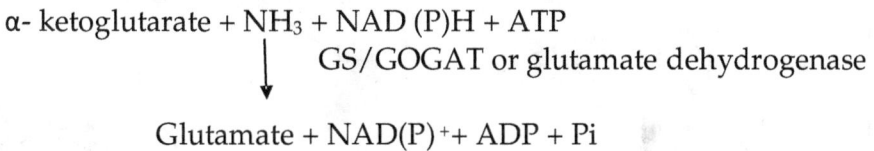

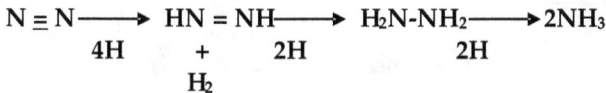

$$\text{Glutamate} + \text{ATP} + \text{NH}_3 \longrightarrow \text{glutamine} + \text{ADP} + Pi$$

Glutamine
Synthetase
(GS)

$$\text{Glutamine} + \alpha-\text{ketoglutarate} + \text{NAD(P)H} \rightleftharpoons 2\,\text{glutamate} + \text{NAD (P)}$$

Glutamate
Synthase
(GOGAT)

$$\alpha\text{- ketoglutarate} + \text{NH}_3 + \text{NAD (P)H} + \text{ATP}$$

GS/GOGAT or glutamate dehydrogenase

$$\text{Glutamate} + \text{NAD(P)}^+ + \text{ADP} + \text{Pi}$$

$$\text{N}\equiv\text{N} \longrightarrow \text{HN}=\text{NH} \longrightarrow \text{H}_2\text{N-NH}_2 \longrightarrow 2\text{NH}_3$$

4H   +   2H   2H
H₂

Overall reaction

$$\text{H}^+ + 8\text{e}^- + \text{N}_2 \longrightarrow 2\text{NH}_3 + \text{H}_2$$

$$18\text{-}24\ \text{ATP} \longrightarrow 18\text{-}24\ (\text{ADP} + \text{Pi})$$

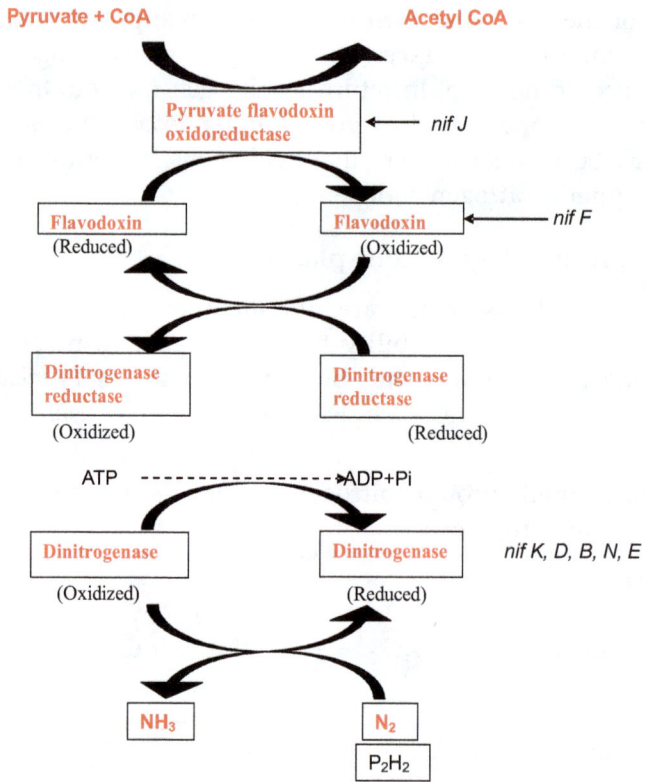

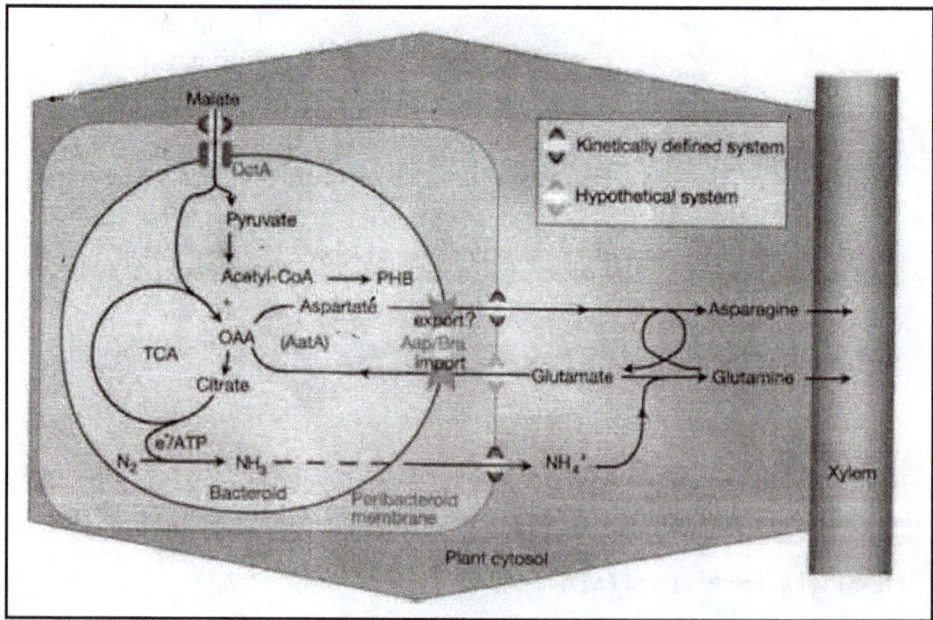

Fig. 9.6. Nitrogen Fixation in legume plants

Biochemistry of nitrogen fixation

Nodules are generally pink in color because of presence of an iron containing substance known as leghaemoglobin. Neither the plant nor the bacterium is individually capable of leghaemoglobin synthesis. The apoprotein globin is encoded by a plant gene, and the synthesis of haeme moiety is under the control of bacterial genes. Throughout the period during which the bacteriods persist, they actively fix atmospheric nitrogen. The reductant and ATP necessary for nitrogen reduction are derived from photosynthates provided by plants. The fixed nitrogen is excreted from the nodules to the plant vascular system as ammonia. About 15-20 moles of ATP are hydrolased per mole of ammonia fixed. It is provided by aerobic respiration within the bacteriods. Ammonia formed reacts with α-ketoglutarate to form glutamate which may be further converted into glutamine. Similarly, aspartate combines with ammonia to form asparagine. Various other products which are synthesized include glutamine, aspartate and ureides like allantonin and allantoic acid; subsequently transported to plant tissues.

Most important plant bacterial interaction is that between legume plants and bacteria of genera *Rhizobium*, *Bradyrhizobium* and *Azorhizobium*. *Azorhizobium* form stem nodules. In the nodules precise oxygen levels are controlled by the oxygen binding protein leghaemoglobin which function as an oxygen buffer cycling between the oxidized ferric ions and reduced ferrous ions. These forms keep free oxygen levels within the nodule at a low but constant level. The ratio of free leghaemoglobin to bound form to oxygen in the root nodule is in the order of 10000:1.

Bacteriods are totally dependent on the plants for supplying them energy sources for nitrogen fixation. The major organic compounds transported across the peri-bacterial membrane are citric acid cycle intermediates, in particular the C_4 acids succinate, malate and fumarate. They are used as electron donors for ATP production and are converted into pyruvate. Ammonia is transported from bacteriod to plant cell and is assimilated to glutamine by glutamine synthetase enzyme by the plant and subsequently transported to plant tissue.

Genetics of nodule formation: Nod genes

Genes directing specific steps in nodulation of a legume by a strain of *Rhizobium* are called *nod genes*. Many Nod genes from different *Rhizobium* species are highly conserved and are borne on large plasmids called *sym plasmid*. Cross inoculation group specificity is controlled by nod genes. The *nod ABC* genes are common to all species of *Rhizobium* and are involved in the production of chitin like molecules, called nod factors, which induce root hair curling and trigger cortical plant cell division. Nod factors consist of a backbone of N-acetyl-glucosamine to which various substituents are linked. *Nif* genes complex regulate the nitrogenase enzyme synthesis.

Nitrogenase

In the fixation process, nitrogen is reduced to ammonia and ammonia is converted to organic form. The reduction process is catalyzed by the enzyme complex called nitrogenase, which consist of two separate proteins called dinitrogenase and dinitrogenase reductase. Dinitrogenase is the Mo-Fe protein, while dinitrogenase reductase is Fe protein. Some nitrogen fixing bacteria can synthesize nitrogenase that lack molybdenum but contain vanadium.

Acid Rain

It is the collective name given to a number of processes which are involved in the deposition of acidic gases from the atmosphere. Unpolluted rain has a pH of about 5.6 due to dissolved carbon dioxide. However in industrial countries, rain water often has a pH between 4.0-4.5. This difference is due to various oxides of nitrogen and sulfur. Sulfur dioxide and nitrous oxide gases are responsible for acid rain whereas nitrous oxide and hydrocarbons produces photochemical smog. Nitrous oxide and chlorofluorocarbons are responsible for depletion of stratospheric ozone layer. The ecological importance of acid rain is still under debate. Many environmentalists blame acid rain for the damage to trees seen in European and North American forests.

Greenhouse Effect

Solar radiations which fall upon earth surface are reflected back to the outer space but some of these radiations are trapped by atmosphere surrounding the earth while other escapes in outer space. The atmosphere acts like a greenhouse, trapping the heat. Humans can only survive on the earth because of greenhouse effect, but it also leads to increase in earth's surface temperature. This is known as global warming. Carbon dioxide and water vapors are responsible for keeping the surface of the earth relatively warm. Humans are contributing to greenhouse effect by releasing a huge amount of carbon dioxide, methane, chlorofluorocarbons, sulfur dioxide, nitrous oxides and other gases into the atmosphere. Although carbon dioxide is the most important greenhouse gas other gases are important too.

Sulfur Cycle

Sulfur is widely distributed in nature in free and combined form. Autotrophic bacteria such as *Thiobacillus* carry out oxidation of sulfur and convert it into sulfates. Sulfate is absorbed by the plant and is converted into organic sulfur compounds. Heterotrophic bacteria such as *Desulfovibrio* carry out degradation process of organic sulfur and convert it into hydrogen sulfide by bacterial oxidation. *Cromatium* help in conversion of hydrogen sulfide into sulfur by bacterial oxidation process and regenerate sulfur in sulfur cycle again. Sulphur enters the soil in the form of plant residues, animal wastes, chemical fertilizers and rain water. The carbon – sulphur ratio (C: S ratio) of the biological waste is in the range of 100:1. The major reserve of the element is in organic form which is mainly present in sulphur containing amino acids like cysteine, cystine and methionine.

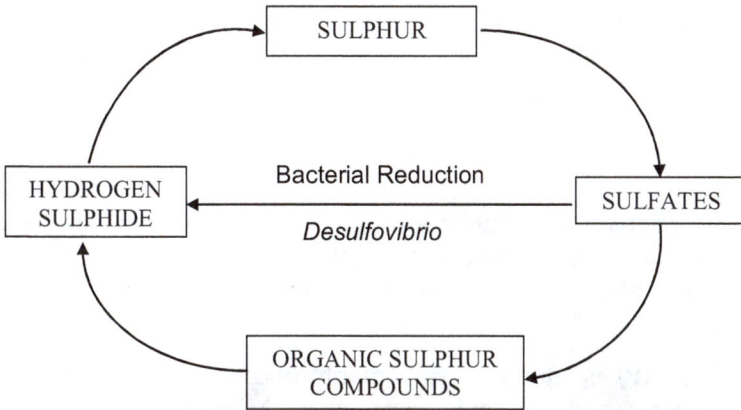

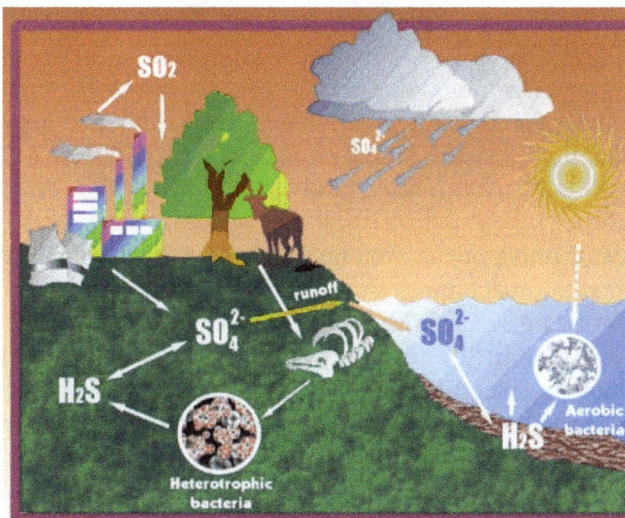

Fig. 9.7. Sulphur Cycle

The sulphur cycle can be divided into four sub stages

(a) **Mineralization** - Decomposition of organic sulphur compounds to inorganic compounds.

(b) **Immobilization** – It is also called microbial assimilation in which a part of mineralized sulphur is utilized for microbial biomass synthesis.

(c) **Sulphur oxidation** - Oxidation of inorganic compounds such as sulphides, thiosulphates and elemental sulphur to sulphate ions. It is caused by variety of bacteria that includes:

- Chemo-autotrophs – *Thiobacillus ferrooxidans*
- Heterotrophic bacteria and actinomycetes – *Beggiata and Thiothrix.*
- Photosynthetic green sulphur bacteria and photosynthetic purple sulphur bacteria.

(d) **Reduction** - Reduction of sulphate and other anions to sulphide e.g. *Desulfovibrio desulfuricans*

Sulphate → Sulphite → Sulfoxylate → Hydrogen sulphide

Sulphide accumulation results in part from sulphate reduction and in part from the mineralization of organic sulphur. When incorporated into the soil, the proteins of plants and animals are hydrolyzed to amino acids. In the aerated environment, the combined sulphur is ultimately metabolized to sulphate. Under water logged conditions, hydrogen sulphide gets accumulated.

Transformation of sulphur is similar to nitrogen. Sulphate dominates the inorganic fraction provided that aeration is adequate. It is taken up by the plant root system largely as sulphate ions.

Phosphorus Cycle

Phosphorus (P) is found in soil, plants and microorganisms and in a number of organic and inorganic compounds. Microorganisms bring about a number of transformations of the element. These include:

(a) Altering the solubility of inorganic compounds of phosphorus.
(b) Mineralizing organic compounds with the release of orthophosphate.
(c) Converting of inorganic available anions into cell protoplasm, an immobilization process similar to nitrogen.
(d) Bringing about an oxidation and reduction of inorganic phosphorus compounds.

Importance of phosphorus cycle in nature is the microbial mineralization and immobilization reactions.

Phosphorus is the second most important mineral nutrient after nitrogen required by both plants and microbes. It plays an important role in accumulation and release of energy during cell metabolism. It is added to the soil as chemical fertilizer or incorporated as plant and animal residues. Agricultural crops contain 0.05 to 0.5 per cent P. It is mainly present in phytin, phospholipids, nucleic acids and nucleoproteins, phosphorylated sugars, coenzymes and related compounds. The organic form of P in soil varies from 25 to 85 per cent of total phosphorus. Microorganisms bring about a number of transformations which help in the phosphorus cycle. These transformations are as follows

(a) Altering the solubility of inorganic compounds of phosphorus.

(b) Mineralizing organic compounds with the release of orthophosphates – Mineralization

(c) Converting the inorganic, available phosphate ions into cell protoplasm – Immobilization

(d) Bringing about an oxidation or reduction of inorganic phosphorus compounds.

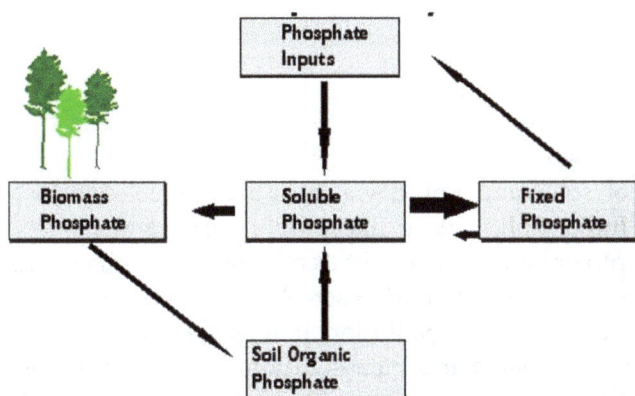

Fig. 9.8. The Phosphorus Cycle

Solublization of inorganic phosphorus

It is caused by production of organic and inorganic acids which lowers the pH of soil. Microorganisms involved in different phosphorus transformation process are following:

(a) Solublization of inorganic P involves microorganisms like *Pseudomonas, Aspergillus, Penicillium, Flavobacterium,* and *Micrococcus*

(b) Mineralization of organic P involves microorganisms like *Aspergillus, Rhizopus, Bacillus* and *Arthrobacter*. Mycorrhizae also play an important role in soil nutrient cycling.

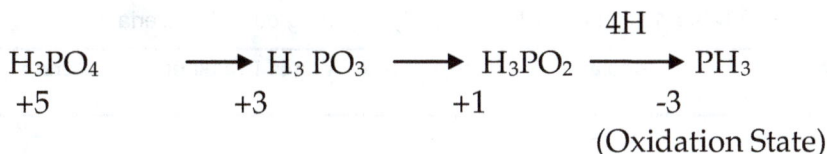

$$H_3PO_4 \xrightarrow{} H_3PO_3 \xrightarrow{} H_3PO_2 \xrightarrow{4H} PH_3$$

$$+5 +3 +1 -3$$

(Oxidation State)

Phosphorus like nitrogen is both immobilized as well as mineralized.

Chapter 10

Microbiology of Waste Disposal and Recycling

Disposal of solid and liquid waste is a big challenge to a developing country like India. Solid waste like crop residue and cattle dung can be decomposed into organic rich and lot of minerals which will enhance nutrient availability to the plants. Liquid waste like sewage disposal is also a big problem as it causes water pollution if it is not properly treated. In this chapter all these problems and their solutions have been discussed in detail.

Sewage Water Microbiology

The principal operation employed in a municipal water purification plant includes sedimentation, filtration and chlorination. Sedimentation occurs in large reservoirs, enhanced by the addition of alum (aluminum sulfate). The water is next passed through sand filter beds, a process which removes 99 per cent of the bacteria, then water is chlorinated:

Water contaminants

The main water contaminants are: Coliform bacteria *Escherichia coli*, *Enterobacter aerogens*. Other enteric groups are *Salmonella, Shigella, Proteus, Pseudomonas* and *Alkaligenes*.

Table 10.1: List of tests identify enteric group of bacteria

Organism	Indole	Methyl red	Voges-Proskauer	Citrate
Enterobacter aerogenes	-	-	+	+
E.coli	+	+	-	-

- Ability to produce indole.
- Amount of acidity produced in glucose broth medium and detected by the indicator methyl red.
- Ability to produce the compound acetyl methyl carbinol in glucose peptone medium –Vi test.
- Utilization of sodium citrate as carbon source.

The routine bacteriological procedures consist of

(a) Plate count to determine the number of bacteria present and

(b) Tests to reveal the presence of coliform bacteria.

Presumptive test

Inoculation in lactose broth is done. If gas is produced, it is presumptive evidence of coliforms. If no gas is produced then coliforms are absent.

Confirmed test

Brilliant green lactose-bile broth (BGLB) medium inhibits growth of lactose fermentors other than coliforms, thus gas formation in BGLB medium constitutes a confirmed test for coliforms.

Eosine-methylene blue agar (EMB) medium

E.coli forms small colonies having dark, almost black centre with greenish metallic shine; *Enterobacter-* forms large, pinkish mucoid colonies, dark centre; rarely show metallic sheen, when grown on eosine-methyl blue agar (EMB) medium.

Completed test

The most typical colonies selected from EMB plates when inoculated into lactose broth, coliforms produce gas. Gram staining of these colonies show Gram negative non-sporulating bacilli.

Measurement of Level of Pollution

Biochemical oxygen demand

Biochemical Oxygen Demand (BOD) refers to the amount of oxygen that would be consumed if all the organics in one litre of water were oxidized by bacteria and protozoa. Microorganisms such as bacteria are responsible for decomposing organic waste. When organic matter such as dead plants, leaves, grass clippings, manure, sewage, or even food waste is present in a water supply, the bacteria will begin the process of breaking down this waste. When this happens, much of the available dissolved oxygen is consumed by aerobic bacteria, robbing other aquatic organisms of the oxygen they need to live. Biological Oxygen Demand (BOD) is a measure of the oxygen used by micro-organisms to decompose this waste. If there is a large quantity of organic waste in the water supply, there will also be a lot of bacteria present working to decompose this waste. In this case, the demand for oxygen will be high so the BOD level will be high. As the waste is consumed or dispersed through the water, BOD levels will begin to decline. Nitrates and phosphates in a body of water can contribute to high BOD levels. Nitrates and phosphates are plant nutrients and can cause plant life and algae to grow quickly. When plants grow quickly, they also die quickly. This contributes to the organic waste in the water, which is then decomposed by bacteria. This results in a high BOD level. When BOD levels are high, dissolved oxygen (DO) levels decrease because the oxygen that is available in the water is being consumed by the bacteria. Since less dissolved oxygen is available in the water, fish and other aquatic organisms may not survive. The first step in measuring BOD is to obtain equal volumes of water from the area to be tested and dilute each specimen with a known volume of distilled water which has been thoroughly shaken to insure oxygen saturation. After this, an oxygen meter is used to determine the concentration of oxygen within one of the vials. The remaining vial is then sealed and placed in darkness and tested five days later. BOD is then determined by subtracting the second meter reading from the first.The range of possible readings can vary considerably: water from an exceptionally clear lake might show a BOD of less than 2 ml/L of water. Raw sewage may give readings in the hundreds and food processing wastes may be in the thousands.

Chemical oxygen demand

The COD (Chemical Oxygen Demand) test represents the amount of chemically digestible organics (food). COD measures all organics that were biochemically digestible as well as all the organics that can be digested by heat and sulfuric acid. It is used in the same applications as BOD. COD has the advantage over BOD in that the analysis can be completed within a few hours whereas BOD requires 5 days. The major drawback of the COD test is the presence of hazardous chemicals and toxic waste disposal. Like the BOD, oxygen is used to oxidize the organics to carbon dioxide and water. However, instead of free dissolved oxygen, chemically bound oxygen in potassium dichromate $K_2Cr_2O_7$ is used to oxidize the organics. As the potassium dichromate is used up, the Cr^{+3} ion is produced. The amount of dichromate used is proportional to the amount of organics present. Likewise, the amount of Cr^{+3} ion present is proportional to the amount of organics digested.

$$\text{Organics} + K_2Cr_2O_7 \longrightarrow Cr^{+3}$$
$$\quad\quad\quad\quad\text{(Orange)}\quad\quad\quad\quad\quad\text{(Green)}$$

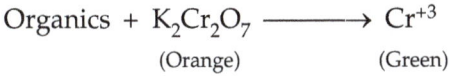

Sewage treatment plant

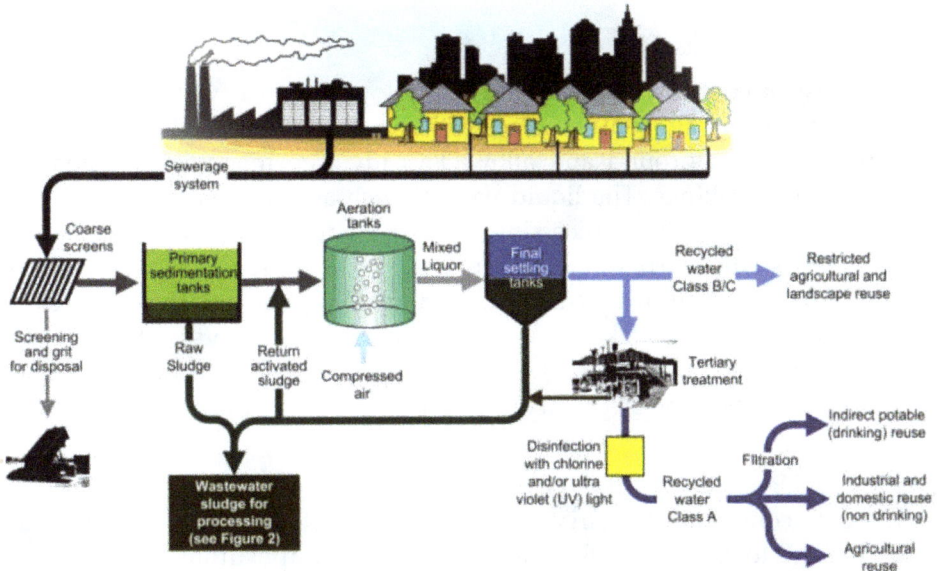

Fig. 10.1. Sewage treatment plant of a Industry

Primary treatment

To remove coarse solid and to accomplish removal of settable solids, sewage water is treated to remove coarse solid materials by mechanical methods like screening, grinding and grit chambers. Sedimentation units (tanks basis) for concentrating and collecting the particulate material referred to as sludge, are used.

Secondary treatment

To oxidize organic constituents of the sewage that is to reduce the BOD, filtration by intermittent sand filters, trickling filters and oxidation ponds is done.

Final treatment

To disinfect and to dispose of liquid effluent and to dispose off solid wastes, the liquid effluent is disinfected by treatment with chlorine.

When populations were small and separated, the amount of waste material available was also small and disposal of such small quantities of wastes was not a problem. However with increase in the population, intensive agriculture and industrialization, the amount of waste has steadily increased. All these factors have created serious problems of disposal. Two major types of wastes are recognized:

- The solid wastes
- The liquid wastes

In most of countries, solid wastes are allowed to decay and decompose in an open area away from the place of habitation. In some other countries, it is buried in the ground and allowed to undergo microbial degradation or used in land fillings. The liquid wastes, such as sewage are drained into the rivers or lakes after processing. In large cities of the West, domestic sewage is usually carried through a well planned sewer system and is collected at a place far away from the place of habitation and then processed and drained into river, lakes or sea. With limiting amounts of fresh water available for both drinking as well as for agricultural purpose, reclaiming of sewage has become necessary.

Sewage Disposal

Sewage constitutes the largest liquid waste in cities and is heterogeneous with regard to both microbial and chemical composition. Generally, it contains about 1-2 per cent solid and 98-99 per cent water. Domestic sewage contains human excreta and hence it is a source of both pathogenic as well as non pathogenic micro-organisms. Organisms commonly found in sewage include the enteric bacteria namely *Salmonella, E. coli, Schigella* and other bacteria like *Pseudomonas, streptococcus* etc.

Sewage from larger cities is also mixed with industrial wastes containing non-biodegradable toxic materials. Because of a high organic mater content, sewage has high biological oxygen demand (BOD). The major purpose of sewage treatment is to reduce this BOD before it is disposed off. BOD is measured by incubating a sample of the liquid at 20°C in sealed container and the amount of oxygen consumed in a definite interval of time (5 days) is measured. A high BOD represents a large amount of degraded organic matter in the sample. Unpolluted natural waters have little or no BOD. A decrease in the BOD during sewage treatment is a reflection of conversions of organic to inorganic materials. The method of sewage purification depends on the BOD level. The term BOD is used to designate the oxygen consuming capacity of a liquid which is a measure of the level of degradable organic matter present.

Oxidation Ponds (Lagoons)

Oxidation ponds (also called lagoons or stabilization ponds) are shallow ponds (2 to 4 ft. in depth) designed to allow algal growth on the sewage effluent. Use of oxidation ponds should be preceded by primary treatment. Oxygen for biochemical oxidation of nutrients is mechanically supplied along with the oxygen supplied by chlorella through photosynthesis.

Primary Treatment	**Secondary Treatment**	**Final Treatment**
(Physical)	(Biological)	(Chemical)

Dry disposal

Activated Sludge Process for sewage treatment

- Normally sewage treatment involves a primary and secondary treatment. In primary treatment large materials such as rags, sticks, trash etc. are removed by passing the sewage through a series of screens and is then allowed to settle for a period of 1-2 hours.

- In some of the sewage treatment plants, aluminum sulphate or ferrous sulphate is added to facilitate coagulation and quick settling.

- The fluid is then transferred for secondary treatment and the sediment is transferred to digesters for further treatment or burnt.

- In secondary treatment the organic matter is stabilized and the BOD is reduced aerobically and this is called as the activated sludge process. During this treatment, the liquid is aerated vigorously and as a consequence it is converted into cell biomass.

- Micro-organisms isolated from the secondary stage are mostly bacteria which include *Pseudomonas, Alcaligenes, Flavobacterium.* These are responsible for bio-degradation of organic load.

- The anaerobic process may reduce the BOD by 35-75 per cent while aerobic activated sludge treatment reduces about 95 per cent of the BOD.

- To remove inorganic chemicals, a tertiary treatment has been used in some of the western countries.

- In this, lime is added to coagulate and precipitate phosphate and allowed to settle. Water is removed by passing the effluent over a stripping tower while the water is broken into small droplets allowing ammonia to escape the water.

- In small sewage plants, a process called the trickling filter method is also used. In this, the sewage is allowed to trickle over coarse sand or pebbles coated with micro-organisms that oxidize the effluents. By adjusting the rate of flow over the filters, efficient oxidation of the sewage can be achieved.

- The activated sludge process is suitable for large communities. For small communities or industries, a process called lagooning which involves storage of the sewage water or effluent in shallow pond or Lagoons to allow settling and stabilization by both aerobic and anaerobic organisms. Pathogenic bacteria are eliminated during this process.

- The treated water can be reutilized either for crop production or for industrial purposes.

Solid Waste Recycling (Composting)

The utilization of human and animal solid wastes is of great importance that is both from the public health and agricultural point of view. Composting is the conversion of the organic fraction of solid waste materials through microbiological processes leading to the production of humus like end product that is used primarily as a soil conditioner. There are basically two methods of composting. In the first method, decomposable material is placed in pits or in stacks on the ground and the material is turned over at intervals to allow adequate aeration and mixing. This method is proved most satisfactory. In the second method, the waste is filled in pits or trenches and allowed to decompose without being turned over.

In big cities where the amount of waste is large, mechanized composting plants exist which handle large quantities of heterogeneous wastes containing both degradable and non-degradable materials. In these plants garbage is first sorted out, the organics are then crushed and heaped to allow decomposition. The quality, composition and nature of the wastes available for composting vary widely with the season and region. In villages, the types of waste depend on the type of agriculture, animals and economic conditions of the area. In cities, the quality and quantity of waste is determined by the waste products of industries, street sweepings, ash and garbage.

Segregation of wastes into compostable and non-compostable material is desirable especially when town wastes which contain metal cans, glass, china pieces, concrete, plastics etc. Metal cans and other ferrous metals can be separated by magnetic separators while other objects can be mechanically removed by screening or by hand picking. Shredding or grinding makes the material more accessible to microbial attack by increase in the surface area. Decomposition of such shredded material occurs at a faster rate. There are several factors which affect the composting process since these affect the microbial activity during decomposition. These include the quality of waste, carbon to nitrogen ratio (C/N ratio), moisture content, pH, temperature, aeration and climatic conditions.

The decomposition of organic matter is brought about by micro-organisms which utilize carbon and nitrogen in the waste for synthesizing cellular constituents. The C/N ratio is, therefore, an important factor determining the rate and extent of decomposition. When excess of carbon available (C/N ratio is high); the rate of decomposition is slow. Generally a C/N ratio at 30-40 is considered as optimum. During decomposition, a part of the carbon is assimilated and a part is oxidized to CO_2 As a consequence, there is a decrease in the C/N ratio and this decrease in the C/N ratio is an important index of composting. Moisture is another important factor that determines successful composting. For satisfactory aerobic composting a moisture content of 40-60 per cent is desirable. If the moisture content is 40 per cent or lower, water can be sprayed during turning. In anaerobic composting (as in the Pit method) the maintenance of moisture content is not necessary since there is not much loss from the initial moisture level. During decomposition due to microbial activity, a large amount of heat is generated. Normally a temperature of 45-50°C is reached in the first 24 hours of composting and of 60-70°C are obtained in 2-5 days. The decline in temperature is gradual.

Microbiology of Compositing

Compostable waste materials normally contain a large number of different types of bacteria, fungi and actinomycetes. During decomposition, changes in the nature and number of these microbes take place. The temperature and availability of nutrients have a great effect on the type of organism present at any time during composting. In aerobic composting, facultative and aerobic fungi, bacteria and actinomycetes are the most active. Mesophilic bacteria and fungi are active in the beginning of composting when the temperature is low. This is followed by thermophilic bacteria and thermotolerent fungi as the temperature of the material increases. During this thermophilic stage, most of the mesophiles are killed. In the terminal stages when the temperature returns to the mesophilic stage, the actinomycetes and mesophilic predominate.

Fungi and actinomycetes play an important role in the decomposition of cellulose and ligin present in composting material. A few representative organisms that predominate at various stage of composting are listed below:

A. Mesophillic Bacteria

- *Cellulomonas folia*
- *Chondrococcus*
- *Myxococcus*

B. Thermotolerent and Thermophilic Bacteria

- *Bacillus streptothermophilus*
- *Actinomycetes*
- *Micromonospora vulgaris*
- *Nocardia brasliensis*
- *Streptomyces rectus*
- *Streptomyces thermovulgaris*

C. Mesophilic Fungi

- *Fusarium roseum*
- *Fusarium culmorum*
- *Copernicus cinereous*
- *Copernicus logopus*
- *Trichoderma viride*
- *Trichoderma lignorum*
- *Rhizopus nigricans*
- *Aspergillus niger*

D. Thermo tolerant andThermophylic Fungi

- *Aspergillus fumigatus*
- *Mucor pusillus*
- *Absidia romosa*

BIOGAS

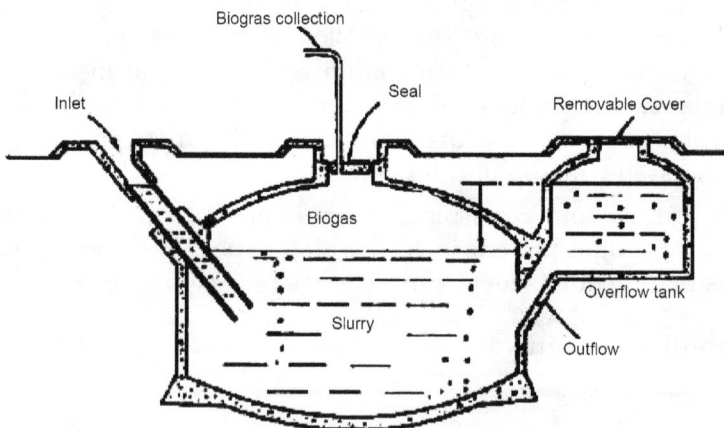

Fig. 10.2: Flow diagram of Solid Waste based Bio Gas Plant

Biogas is mixture of methane 50-60 per cent, carbon dioxide 30-40 per cent, hydrogen 5-10 per cent, H_2S and nitrogen (traces) produced from the anaerobic digestion of animal or plant wastes or any cellulose containing waste material. The digester used for biogas production is called a biogas plant. A typical biogas plant using cow dung as raw material consists of digester and gasholder. The digester is either of batch type which is filled once, sealed and emptied when the raw material stop producing gas or of continuous type which is fed with a definite quantity or waste at regular intervals so that gas production is continuous and regular. The nature of fermentation in the digester is anaerobic.

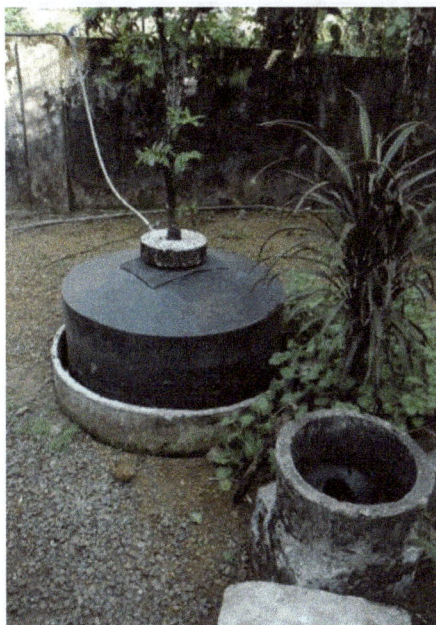

At present four different bacterial groups are recognized to be involved in the anaerobic fermentation of organic matter to methane.

- The hydrolytic bacteria which catabolise carbohydrates, proteins, lipids other component of biomass to fatty acids, H_2 and CO_2.

- The H_2 producing acetogenic bacteria which catabolise certain fatty acid to acetate, CO_2 and H_2.

- The homoacetogenic bacteria which synthesize acetate using H_2, CO_2 and formate or hydrolyze multi carbon compounds to acetic acid.

- The methanogenic bacteria which utilize acetate, CO_2 and H_2 to produce methane.

The first groups of bacteria include facultative as well as strict anaerobes like *Cellulomonas, Clostridium, Bacillus, Bactoroides, Rumenococcus, Eubacterium* etc.

The second group of bacteria include *Acetobacter* and *Acetococcus*.

Table 10.2 : Third Group include major genera of *Methanogenic* Bacteria

Genus	Morphology
Methanobacterium	Long rods or filament
Methanomicrobium	Short rods
Methanogenium	Irregular
Methanococcus	Small circular
Methanospirillum	Short to long spiral
Methanosarcina	Circular

Optimal pH for methane production is 6.8-7.2. If pH drops to 6.6 or below, there is an inhibition of *methanogenesis*.

Chapter 11

Biofertilizers: Low-Cost and Eco-friendly Strategy

Agricultural productivity in Indian subcontinent has gained encouraging trends during last four decades. High yielding variety seeds, availability of more water for irrigation and enhanced use of chemical fertilizers have been the main factors for achieving the high productivity. However, the pathway adopted by us has been dependent on non-renewable energy resources, resulting in an exponential increase in the consumption of petroleum products. Urea is the main fertilizer being used across the globe in maximum quantities as compared to any other fertilizer. All the urea manufacturing units depend upon petroleum products. According to an estimate, the manufacture, transportation and application of one 1.0 kg urea involve an expenditure of 1.0 kg petroleum products. Besides, an excessive use of urea to supplement nitrogen to the soil may render the groundwater polluted. Nitrate pollution in water may cause awful diseases like methaemoglobinemia and hypertension among the infants rendering them handicapped. In other words, excessive use of urea is not only expensive but unsafe also for human health and environment.

In view of sky rocketing population and growing grain demand, the necessity of intensive agriculture is likely to continue. Regular replenishment of plant nutrients to maintain the soil fertility is unavoidable. Consequently, any curtailment in the consumption of urea and other chemical fertilizers would not be feasible. In view of necessity of intensive agriculture and keeping economy, health and environment in mind; the need of the hour is to exploit all possible sources of plant nutrients so as to achieve required productivity through intensive agriculture. Agriculturists suggest that the requirements of plant nutrients can be fulfilled only when the chemical fertilizers are judiciously used along with green manure, organic manure and biofertilizers.

Biofertilizers are environment friendly, highly efficient and low cost agricultural inputs. The use of biofertilizers for various crops is directly or indirectly, a true service to the soils of nation and the environment.

What are Biofertilizers?

These are preparations carrying live or latent microorganisms having characteristic capability of fixing atmospheric nitrogen, solubilizing unavailable phosphate or composting the agricultural waste. Biofertilizers may be applied to the soil through seeds, roots and directly to soil where microbes multiply and mobilize the inert nutrients.

Types of Biofertilizers: Biofertilizers can be generally categorized into three types

(a) Nitrogen Supplementing

(b) Phosphate solubilizing

(c) Composting microorganisms.

Nitrogen Supplementing Microorganisms

These microorganisms have the capability of fixing atmospheric nitrogen which is 78 per cent of the atmosphere. Most of the plants can utilize nitrogen only in the form of nitrate. Hence unless the nitrogen gas is converted to nitrate, it remains unavailable for plants. Certain microorganisms absorb nitrogen gas as their feed and convert it into ammonia through the activity of an enzyme called nitrogenase. Ammonia is converted into nitrate by nitrification or directly assimilated into the plant system.

Depending upon the requirements of various types of crops, the following microorganisms are being manufactured on commercial scale as biofertilizers.

• *Rhizobium*

This bacterium fixes atmospheric nitrogen in the symbiotic association with the leguminous crops. *Rhizobium* enters the root system after germination of seeds and nodules are developed on the roots. These nodules inhabit *rhizobia*, which fix atmospheric nitrogen and keep supplying ammonia to the plant. *Rhizobia* are host specific as they form nodules and fix nitrogen on specific hosts. Hence, while procuring *Rhizobium* culture, it should be taken care of that name of the pulse crop should be mentioned on the culture for which it is used.

Benefited Crops: Soybean, groundnut, berseem, sesbania and all other pulse crops.

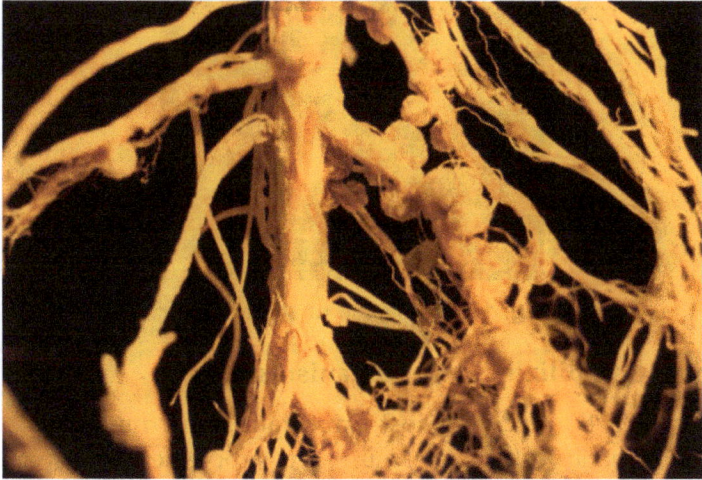

Fig. 11.1. Nodules shown on the roots of legume plant

• *Azotobacter*

These bacteria fix atmospheric nitrogen in free living conditions. They multiply in the vicinity of the root system and convert atmospheric nitrogen to ammonia. Plants assimilate the fixed nitrogen. The capability of fixing nitrogen in free living conditions is accredited to *Azotobacter* as a versatile biofertilizer which can be successfully utilized against a broad range of crops belonging to different groups for supplementing chemical nitrogen. In addition to fixing nitrogen, they also produce plant growth regulating substances in the vicinity of the plant.

Benefited crops: Wheat, maize, sorghum, pearl millet, mustard, sunflower, cotton, fruits and flowers yielding crops, tea, coffee, vegetables etc.

• *Acetobacter*

Similar to *Azotobacter*, this bacteria fixes nitrogen in aerial as well as underground parts of the plant. Most common species is *A. diazotrophicus* fixing nitrogen in sugarcane. Various field studies revealed that *Acetobacter* works more efficiently for sugar yielding crops like sugarcane and sugar beet. It has been estimated that approximately one-fourth of total nitrogen requirement of sugar yielding crops can be fulfilled by this bacteria. These bacteria are endosymbiont as they remain within the plant.

Benefited crops: Sugarcane, sugar beet and pearl millet.

Fig. 11.2. A glance of *Acetobactor*

• Phosphate Solubilizing Microorganisms (PSM)

Phosphate is the second most important plant nutrient. In general, chemical phosphatic fertilizers are used to supplement phosphates to the soil. Scientific experiments have proved that 30 to 35 per cent of phosphatic fertilizers applied are actually utilized by the plants while remaining 65 to 70 per cent of chemical phosphatic fertilizer change to insoluble state and become unavailable to the plants. Certain microorganisms have the capability of resolubilizing this insoluble phosphate, making it available to the plants. PSM is a balanced blend of certain efficient phosphate solubilizing microorganisms which work under diverse geographical conditions. Since PSM has the capability of working in various types of soils under free living conditions, this biofertilizer can be utilized against all the crops with equal efficiency e.g. *Bacillus*, *Pseudomonas* and *Aspergillus*.

•Composting Microorganisms

The use of compost and farm yard manure to replenish the nutrients in the soil is prevailing since ancient times. Dead leaves, plant parts and other agricultural trash has got sufficient plant nutrients, but these are unavailable to crop plants unless their complex form is changed to simpler form through microbial decomposition. This process of decomposition is known as composting and involves specific microorganisms. Composting microorganisms are available in the atmosphere and continue decomposing the dead organic matter. In case the population of efficient composing microorganisms is increased over the heap of agriculture waste, the process of composting becomes faster and a good quality compost or organic manure is prepared in merely one-fourth time as compared to natural composting. The organic manure so obtained carries almost all the required plant nutrients in balanced quantities. The organic manure preparation can be fastened by the use of *Trichoderma*, *Penicillium* and *Aspergillus*.

• **Urea Coating Agents (UCA)**

Nitrogen deficiency in soil is generally replenished by application of urea, but approximately 30 per cent is actually utilized by the plants; while the rest 70 per cent either leaches down to ground water or volatilized back to atmosphere. Immediately, after its application urea tends to break into nitrates. This process is known as nitrification which is much quicker than the nitrate assimilation by the plants. Consequently about 70 per cent of urea goes waste and causes pollution.

Urea Coating Agent (UCA) is a balanced blend of certain herbs and minerals which inhibits the process of nitrification, resulting in slow release and more assimilation of urea by the plants. It is estimated that 40 to 50 per cent saving of urea can be achieved by coating the urea granules before application.

• **Biofungicide**

We are aware of the losses due to certain fungal diseases in various crops. Generally, chemical fungicides are used to combat the fungal diseases. These poisonous chemicals persist in the environment for a long time and impose a slow but harmful effect on living beings and ultimately on human health.

Table 11.1: Doses of various biofertilizers for different crops

Sr. No.	Target Crops	Seed Treatment (g/kg)		Soil Application (kg/ha)	
		Trichoderma	*Acetobector*	*Trichoderma*	*Acetobector*
1.	All pulses crops Soybean, Groundnut, Mung, Urd, Lentil, Pea, Gram etc.	4-5	--	2.5	--
2.	A. Cereals, millets oilseed, wheat, jowar, bajra, Mustard and Sunflower etc.	4-5	--	2.5	--
	B. Cash crops (Sugarcane, Potato) Vegetables & Fruit crops.	4-5	--	2.5	--
3.	Sugarcane & Sweet potato	--	2.5 kg	2.5 kg	2.5 kg

*1. U.C.A. -1 kg/50 kg urea bag
*2. Composting Culture —1 kg for 2-3 metric tonnne of agricultural waste

Trichoderma is a specific fungus having characteristic capability of inhibiting the growth of a broad range of pathogenic fungal species. Due to being biological, this biofungicide has got no adverse effect on the environment. Application of *Trichoderma* is known to prevent various diseases like stem and root rot, damping off, wilt, blight and other diseases of leaves. The other microorganisms which have been successfully used as biopesticide in different food crops include *Bacillus* and *Pseudomonas* species. Some of the experiments conducted by our laboratory had shown that *Pseudomonas maltophilia* strain PM4 can be used as bioinsecticide against white grub in groundnut, for aphid control in mustard and for termite control in wheat.

Table 11.2: Doses of various biofertilizers for different crops

Sr.No.	Target Crops	Seed Treatment			Soil Application		
		Rhiozo-bium	*Azoto-bacter*	PSM	*Rhiozo-bium*	*Azoto-bacter*	PSB
1.	All pulses crops like Soybean, Groundnut Mung, Urd, Lentil, Pea Gram etc.	50ml/10 kg seed	--	50ml/10 Kg seed	1.5 litre	--	
2.	A. Cereals, Millets oilseed, Wheat, Jowar, Bajra, Mustard and Sunflower etc.	--	50ml/10 kg seed For large seed crop And 50ml for small seed crop	-Do-	--	2 litre liquid culture	2 litre liquid culture
3.	B. Cash crops {Sugarcane , Potato, Vegetables & Fruit crops}	--	1.5 litre	1.5 litre liquid culture	--	2 litre	3 litre

Chapter 12

Symbiotic Associations among Microorganisms

The three beneficial relationships are symbiosis, protocooperation and commensalism. Sometimes the benefit is mutual, but commensal relationships are quite frequent. For example, cellulolytic fungi produce from cellulose a number of organic acids that serve as carbon sources for non-cellulolytic bacteria and fungi. A second type of commensal association arises from the need of many microorganisms for growth factors. These compounds are synthesized by certain microorganisms, and their excretion permits the proliferation of nutritionally fastidious soil inhabitants.

Symbiotic associations are evident in soil among several groups of organisms: algal and fungi in lichens, bacteria residing within protozoan cells, bacteria and roots in the *Rhizobium* – legume symbiosis, fungi and roots in mycrorrhizae, protozoa in underground termites.

Microbial Competition

The categories of deleterious interactions are summarized by the terms competition, amensalism, parasitism and predation. Various factors responsible for competition include the rivalry for limiting nutrients or other common needs or the release by one species of toxic substances, or the direct feeding of one organism on a second. Competition for carbon, inorganic nutrients or O_2 is quite common. Predation and parasitism are observed in the feeding on bacteria by protozoa, the attack on nematodes by predacious fungi, the digestion of fungal hyphae by bacteria and the lyses of bacteria and actionomycetes by bacteriophages. Some microbes inhibit the growth of other microorganisms by producing some secondary metabolites in the environment; these metabolites are known as antibiotics. An antibiotic is a substance formed by one organism in a low concentration that inhibits the growth of another organism of unrelated species. A variety of actionomycetes, bacteria and fungi are able to synthesize antibiotics. Actinomycetes are particularly active in this regard and streptomycin, chloramphenicol, cycloheximide, cycloheximite, chlortetracycline are some of the important chemotherapeutic substances synthesized by them. Antibiosis is especially common among *streptomyctes* isolates, but *Nocardia* and *Micromonospora* are also active. *Bacillus* and *Pseudomonas* also liberate antibiotic substances into its surroundings. Species of *Penicillium, Trichoderma, Aspergillus, Fusarium* and other fungal species also excrete antibiotics.

Some bacterial strains inhibit the growth of other strains of the same species by producing metabolites; these metabolites are known as bacteriocins. Hence bacteriocins differ with the antibiotic as they inhibit growth of other strains of the same species while antibiotics inhibit growth and other species.

There is wide variety of relationships that exists between different types of microorganisms. Some of the common interactions which are prominent in nature are as follows:

Neutralism: An association, in which two microorganisms of different species behave independently, is called neutralism.

Symbiosis: An association, in which two symbionts relying upon one another and both are benefitted by the relationship, is called symbiosis e.g. an association between legume plant and *Rhizobium*. *Rhizobium* fixes atmospheric nitrogen to the plants and plants in turn provide shelter and food to them.

Proto-cooperation: An association of mutual benefit to the two species but without cooperation being obligatory for their existence or for their performance reaction e.g. synergism between phosphate solubilizers and plants and crop.

Commensalism: An association in which one species drives benefit from the company of its associate while the other is unaffected.

Competition: A condition in which there is suppression of one organism as two species struggle for limiting factors like oxygen, nutrients, space or other common requirements.

Ammensalism: In which the growth of one species is suppressed due to production of toxins like antibiotics, harmful gaseous compounds like NH_3, CO_2, ethylene, nitrite, HCN etc. while the second is unaffected. Growth of *Nitrobactor* and fungi may be affected adversely by the large amount of NH_3 released during decomposition of leguminous green manures.

Parasitism or Predation: An association in where one partner lives in or on the body of the other (host) and feeds on the body fluids of the host. Earthworms and other macroorganisms are prone to parasitic attack by bacteria, fungi, viruses etc.

MYCORRHIZAE

It is a Greek word made up of two words; mycofungus and rhiza- root. It is a symbiotic association between plant roots and fungal mycelia. Two primary types of mycorrhizal associations occur, that are ectotrophic or ectomycorrhiza and the endotrophic or endomycorrhiza.

The ectomycorrhizal association primarily exists in the families of *Pinaceae, Fagaceae, Soliaceae* etc. The fungal partner belongs to Basidiomycetes and Ascomycetes. In ectomycorrhizal association, the fungal mycelium completely encloses each feeder rootlet in a sheath or mantle of fungal hyphae and hyphal branches penetrate space between cells of the root cortex (intercellular). These fungi produce plant growth substances that induce morphological alteration in the roots, causing short dichotomously branched mycorrhizal roots.

Endomycorrhizae are prevalent in most families of angiosperms, in certain pteridophytes and bryophytes. They are also known as vesicular arbuscular mycorrhiza (VAM), which possesses special structures known as vesicles and arbuscles, the latter helping in the transfer of nutrients from soil to the root system. These fungi are classified on the basis of spore morphology into six genera. *Glomus* is most important of them. In VAM, fungi are intracellular and obligate endosymbionts.

Physiology of VAM/ Endomycorrhizae

They help plants in absorption of nutrients and water. The surface area is increased and thus extending the zone of nutrient absorption for poorly mobile elements such as Cu, P, Zn, S and N.

Physiology of ectomycorrhizae

They do not secrete cellulolytic and lignolytic enzymes and depend upon the plant host for carbon nutrients. Auxin produced by fungi causes dichotomy of rootlets. These rootlets generally lack root hairs. They form Harting Net or mantle which helps in absorption of nutrients. About 80-90 per cent of absorbed phosphorus remains in fungal sheath. They also protect roots from fungal pathogens.

Chapter 13

Microbial Fermentation

Fermentation is an anaerobic (without oxygen) cellular process in which organic foods are converted into simpler compounds, and chemical energy (ATP) is produced. Fermentation occurs in fruits, bacteria, yeasts, fungi and in mammalian muscles. Yeasts were discovered to cause fermentation of sugars as observed by the French chemist, Louis Pasteur. Pasteur originally defined fermentation as respiration without air. However, fermentation does not have to always occur in anaerobic condition. Yeasts still prefer to undergo fermentation to process organic compounds and generate ATP even in the presence of oxygen. However, in mammalian muscles, they turn from oxidative phosphorylation (of cellular respiration) to fermentation when oxygen supply becomes limited, especially during a strenuous activity such as intensive exercising. Fermentation is believed to have been the primary means of energy production in primitive organisms when oxygen was low in the atmosphere, and thus represents a more ancient form of energy production in cells.

Fermentation occurs naturally but humans have used and controlled the process. It is used in the production of *alcohol*, bread, vinegar, and other food or industrial products. Major classes of products and processes of microbial fermentation are as follows

1. Pharmaceutical chemicals: Most prominent in this category are the antibiotics and steroids but other substances such as insulin and interferons are now being produced by genetically engineered bacteria.

2. Commercially valuable chemicals: Solvent, enzymes and intermediate compounds used for commercial purposes.

3. Food supplements: Mass production of yeast, bacteria and algae from media containing inorganic nitrogen salts and other easily available cheap nutrient supplements. Large scale production of amino acids is an attractive industrial process being employed in many parts of the world.

4. Alcoholic beverages: Brewing (beer), wine and production of alcoholic beverages.

5. Vaccines: Some microorganisms are grown in very large scale for use as vaccine. The whole cell or product of the cell is used for preparation of vaccine.

PRODUCTION OF VINEGAR

Various steps used in vinegar production are as follows
1. Build up of yeast inoculum.
2. Alcoholic fermentation of grape juice, apple juice or sugarcane juice to get alcohol up to 10-13 per cent.
3. Exposure of alcohol to acetic acid bacteria.
4. The final product contains about four per cent acetic acid.

For the production of vinegar, Fringes method is generally adopted. Alcohol produced by yeast fermentation of any fruit juice is adjusted to 10-13 per cent alcohol which is acidified with acetic acid and nutrients. Acetic acid bacteria (*Acetobacter*) are inoculated into beech wood shavings. The alcoholic mixture is applied in a trough at the top of the beech wood shavings chamber. As the alcohol passes over the shavings, *Acetobacter* present on the shavings oxidizes some of the alcohol to acetic acid. The mixture is collected at the bottom of the unit and is re-circulated over the shavings resulting in more oxidation of alcohol until the vinegar of desired strength is obtained.

$$2C_2H_5OH + 2O_2 \longrightarrow 2CH_3COOH + 2H_2O$$

Ethanol   Acetic acid

Precautions

- Sufficient supply of air should be provided.

- Temperature should be between 15 and 34°C.

Feeding of alcohol produced from fruit juice by yeast fermentation

Alcohol tank

Holes

Wood shavings onto which *Acetobacter* immobilized

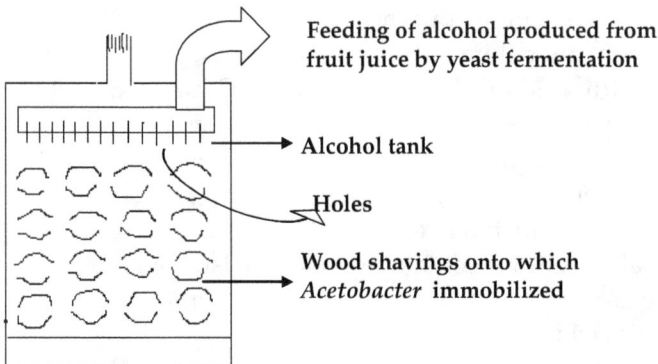

Fig. 13.1. Beach wood chamber with wood shavings (Vinegar production)

PRODUCTION OF LACTIC ACID

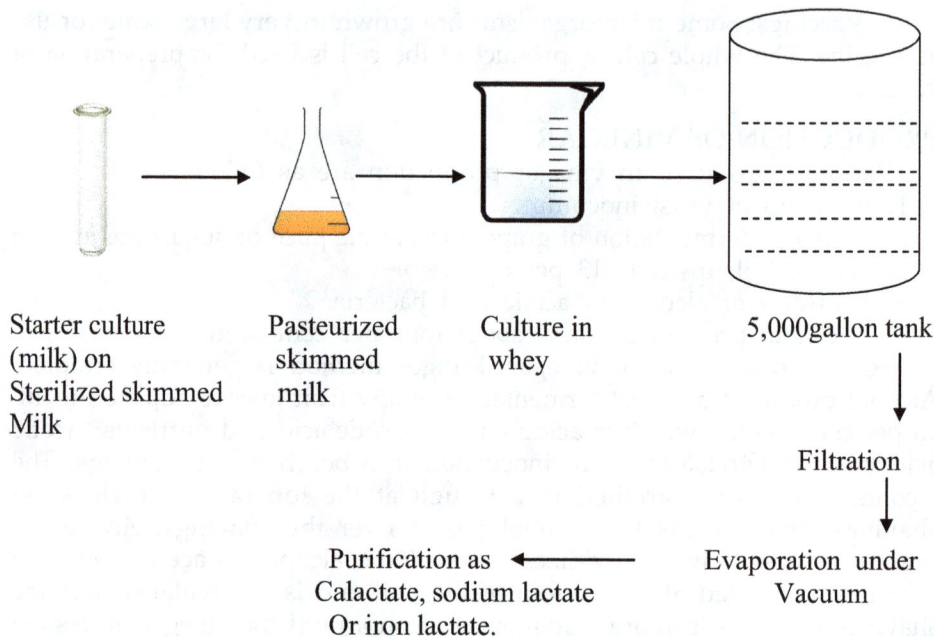

Starter culture (milk) on Sterilized skimmed Milk

Pasteurized skimmed milk

Culture in whey

5,000gallon tank

Filtration

Purification as ← Calactate, sodium lactate Or iron lactate.

Evaporation under Vacuum

Fig. 13.2. Lactic Acid Production

It can be produced from whey using *Lactobacillus bulgaricus*

Reaction

$$C_{12} H_{22} O_{11} + 2O_2 \longrightarrow 2C_6 H_{12} O_6 \longrightarrow 2CH_3 \text{ CHOH COOH}$$

Lactose   Glucose   Lactic acid

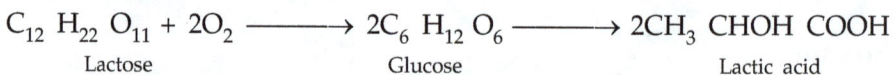

Steps involved in lactic acid production are as follows:

• Preparation of starter culture
• Preparation of inoculum
• Buildup of inoculum in whey.
• Transfer of inoculum into fermenter.
• Temperature should be around 42°C to inhibit the growth of undesirable organisms.
• Addition of lime slurry to neutralize acid produced during fermentation.
• Boiling of the material to coagulate protein
• Removal of filtrate containing calcium lactate.
• Lactate is concentrated by vacuum evaporation.

Uses of Lactic acid

• Removes calcium deficiency.
• Used for anemia treatment.
• Used as plasticizer.

PRODUCTION OF ANTIBIOTICS ON LARGE SCALE

1. Media ingredients for growth of *Pennicillium chrysogenum*

   a. Corn steep liquor

   b. Lactose

   c. Salts

   d. Other growth factors

   Steps involved in production of penicillin are:

   • Mixing of media ingredients.

   • Sterilization of media ingredients.

   • Build up of *Penicillium chrysogenum* inoculum.

   • Inoculation of fermenter vessel with inoculum.

   • Incubation for 4-5 days with continuous aeration.

   • Removal of *Penicillium chrysogenum* by filteration.

   • Removal of penicillin from culture filtrate by precipitation, redissolving and filtration; vaccum drying and packing of antibiotics in injection viols.

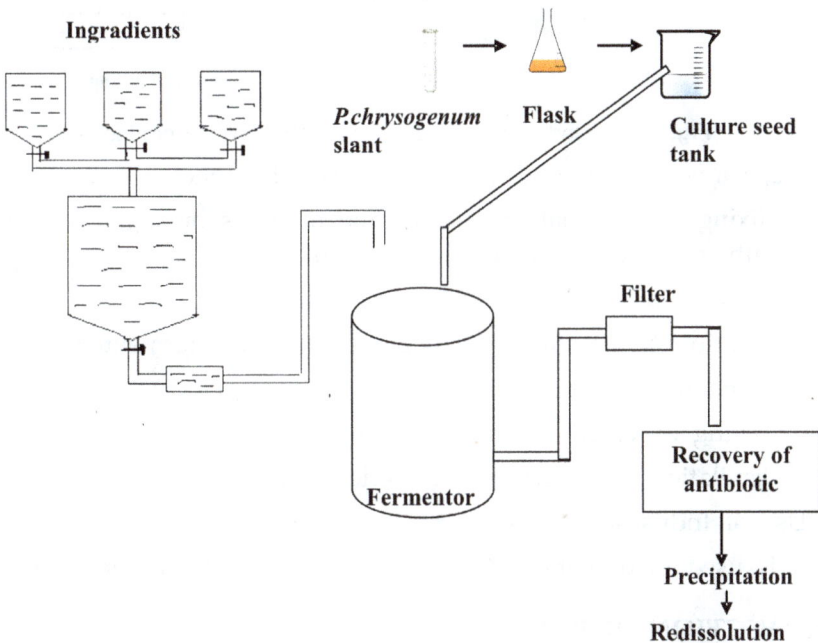

Fig. 13.3. Production of Antibiotics on Large Scale

Uses

Penicillin is used in curing diseases caused by Gram + ve pathogenic bacteria.

PRODUCTION OF INDUSTRIAL ALCOHOL- WINE PRODUCTION

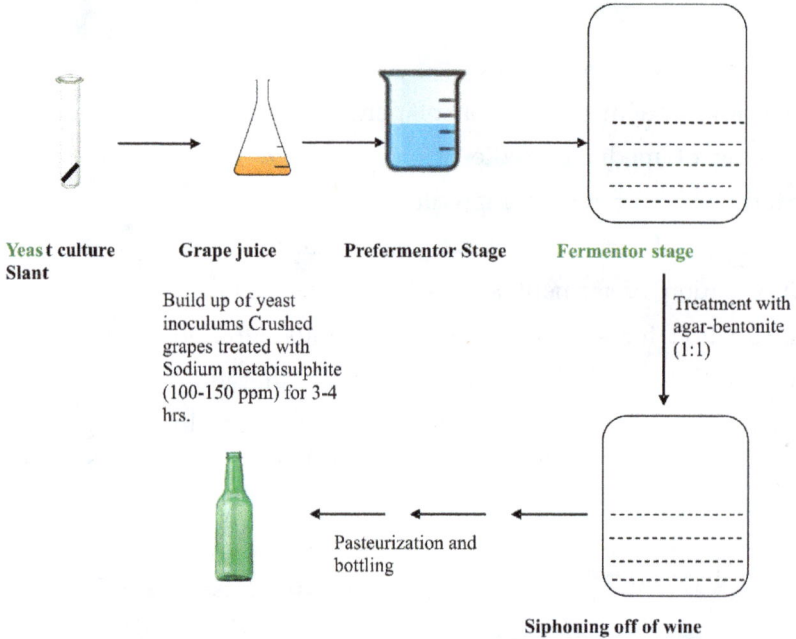

Fig. 13.4. Production of industrial Alcohol : Wine production

Steps involved in production of industrial alcohol

- Mixing of raw material (sugarcane mollases or sugarbeat mollases) with medium ingredient in fermentor.
- Build up of yeast inoculum.
- Addition of inoculums (5-10 per cent) to the fermentor.
- Fermentation is completed within 2-3 days.
- Settling of yeast culture at the bottom.
- Distillation of supernatent to collect alcohol.

Use of Industrial Alcohol

It is used in preparation of rum, whisky and for other industries.

PRODUCTION OF BAKER'S YEAST

Steps involved in production of baker's yeast:

- Preparation of medium for growth of yeast.

- Build up of yeast inoculums.
- Addition of yeast inoculums into mash in fermentor for large scale production of yeast.
- Centrifugation of fermentor contents.
- Collection of yeast.
- Passing through filter press to remove moisture.

Uses of yeast

In bread making, idli, dosa, jalebi etc.

Table 13.4: List of various industrial products, organisms and its uses.

Product	Substrate	Organism	Use
Penicillin	Corn steep liquor, lactose, inorganic salts.	*Penicillum chrysogenum*	Control bacterial infections especially caused by Gram positive bacteria
Streptomycin	Corn steep liquor, sucrose, inorganic salts	*Streptomyces griesus*	Control bacterial infections
STEROIDS Hydroxyprogestron	Ordinary growth medium	*Rhizopus arrhizus*	Life saving drugs and growth harmone
Gibbereillic acid	Ordinary growth medium	*Fusarium moniliforme*	Plant harmone help in fruit setting, seed setting
Indole acetic acid (IAA)	Ordinary growth medium	Many fungal and bacterial species	Helps in seed germination and plant growth.
Citric acid	Cheap carbon source such as cane molasses, sugarcane juice etc.	*Aspergillus niger*	Medicinal value: blood transfusion, Industrial uses: plastic industry and in textile
Lactic acid	Whey , ordinary carbon source	*Lactobacillus* spp.	Food and pharmaceutical Industry
Gluconic acid	Whey , ordinary carbon source	*Aspergillus niger*	Textile, leather and photography
Proteases	Wheat bran and other cheap carbon source	*Rhizopus oryzae, Bacillus* sp.	Removal of laundry spot, tenderization of meat.
Pectinase	Cheap carbon source	*Aspergillus wentii*	Clarifying agents

Contd...

Table 13.4. *Contd...*

Product	Substrate	Organism	Use
Amylases	Cheap carbon source	*Aspergillus* spp.	Converts starch into sugar
Cellulase	Cheap carbon source	*Trichoderma viridae*	Converts cellulose into sugar
Baker's yeast	Wheat bran	*Saccharomyces cerevisiae*	Baking and brewing industry
Soyasouce	Soyabean	*S. rouxii*	Food condiments
Microbial protein	Waste material	*Candida utilis*	Animal food supplement and used as single cell protein (SCP)
Wine	Grape or other fruit juice	*S. cerevisiae*	Contain about 10% alcohol and improves digestion
Whisky	Grain malt	*S. cerevisiae*	Contain 40-50% alcohol, small quantity tone up heart and circulatory system while heavy dose affects vital organ of body
Rum	Sugarcane molasses	*S. cerevisiae*	Contain 40-50% alcohol, inferior from whisky because it contains contaminants.
Beer	Barley malts	*S.cerevisiae*	Contains 6-8% alcohol
Bread	Starchy flour	*S.cerevisiae*	Used as food
Riboflavin	Ordinary growth medium	*Eremothecium ashbyii*	Vitamin supplement
Vitamin B12	Ordinary growth medium	*Eremothecium ashbyii, Propionibacterium*	Food and feed supplement
Solvent Acetone-butanol	Ordinary growth medium	*Citro acetobutyricum*	Industrial and laboratory uses
Ketogluconic acid	Ordinary growth medium	*Gluconobacter suboxidans*	Fine chemical intermediate
Acetic acid	Medium containing 10-12% alcohol	*Acetobacter* spp	Used in pickles having medicinal value
Dextran	Ordinary growth medium	*Leuconstoc mesenteroids*	Blood plasma substitute, stabilizer in food
Glutamic acid	Ordinary growth medium	*Brevibacterium* sp.	Food additives
Lysine	Ordinary growth medium	*Micrococcus glutamicum*	Amino acid supplement
Streptokinase	Ordinary growth medium	*Streptococcus equisimilis*	Dissolve blood clot
Bio-insecticide	Ordinary growth medium	*Bacillus thuringenesis and B. papillae*	Control of insects in plants
Insulin	Ordinary growth medium	*E. coli*	Used for curing diabetes patients
Interferon	Ordinary growth medium	*E. coli*	For cancer treatment
Somatostatin	Ordinary growth medium	*E. coli*	Growth hormone

Chapter 14

Food Microbiology

Microorganisms are associated in a variety of ways with all of the foods we eat. Microorganisms may influence the quality, availability and quantity of our food. Food can serve as a medium for growth of microorganisms and this growth may cause the food to undergo decomposition and spoilage. Some microorganisms, if allowed to grow in certain food products, produce toxin substances that result in food poisoning when the food is ingested. Still other microorganisms are used in the preparation and preservation of food products such as yoghurt and sauerkraut. Hence, the food microbiology deals with the following:

1. Factors that determine microbial growth in foods
2. Role of microbes in food preparation
3. Fermented foods
4. Microorganism as a food – single cell protein
5. Microbial flora of fresh foods
6. Microbial spoilage of food and food poisoning
7. Preservation of foods.

Factors that determining microbial growth in foods

A variety of factors such as the pH, moisture content, oxidation-reduction potential, nutrient etc. influence microbial activity in foods. For example, each organism has an optimum pH for growth. In general, yeast and fungi are more acid tolerant than bacteria. Most yeasts favor a pH around 4- 4.5, while most bacteria favor a pH around 7.0. The pH of foods varies as some may be neutral while others may be acidic. Acidic foods are not readily spoiled by bacteria. Microorganisms have an absolute demand for water and the optimum level of moisture required for growth varies with the organism. Nutrients in food, their kind and proportions determine the type of organism that will grow. Microorganisms vary in their ability to use nutrients. The presence of easily utilizable nutrients will encourage faster growth and quicker spoilage.

Microbes important in foods

Some molds are important in food preparation as molds are involved in ripening of cheese and preparation of oriental foods. Yeasts are both useful as well as problem organism and this depends upon the food. For example, the production of wine is dependent on growth and activity of the yeast *Saccharomyces cerevisae;* while wine can be oxidized to CO_2 and water by wild yeast.

Fermented foods

There are many useful applications of microorganism in food industry. A variety of important products have been mentioned in Table 14.1.

Table 14.1: Various food products prepared by use of microbes

Food	Raw material used	Microbes involved
Curd	Milk	*Lactobacillus, Streptococcus.*
Ripening of cheese.	Milk	*Lactobacillus, Propionibacterium*
Bread	Flour	*Saccharomyces*
Sauerkraut	Cabbage	*Lactobacillus, Leuconostoc, Aerobacter.*
Wine	Fruit juices	*Saccharomyces*
Vinegar	Alcohol	*Acetobacter*
Dosa & Idli	Rice urd daal	*Saccharomyces*
Soysauce	Soyabean	*Aspergillus, Lactobacillus, Saccharomyces*

Single Cell Protein (SCP)

Bacteria, yeasts and algae, produced in massive quantities are attractive sources of food for animals as well as humans. These microorganisms can be cultivated on industrial wastes or by-products as nutrient and yield as large cell crop that is rich in protein.

- Bacterial cells grown on hydrocarbon wastes from the petroleum industry are a source of protein in France.
- Yeast cell crops harvested from the vats used to produce alcoholic

beverages have been used as a food supplement for generations.

Characteristics of these processes are

(i) Microorganisms grow very rapidly and produce a high yield. Algae grown in ponds can produce 20 tons (dry wt.) of protein per acre per year.

(ii) The protein of selected microorganisms contains all essential amino acids.

(iii) Some microorganism particularly yeasts have a high vitamin content.

(iv) The medium (nutrients) for growth of microorganism may contain industrial wastes or by-products.

They are the dried cells of microorganisms such as algae, certain bacteria, yeast and molds and that are grown in large scale culture systems for use as protein for human or animal consumption. The products also contain other nutrients like fat, vitamin and minerals. Various examples are as follows:

1.   Yeast cell in leavened bread.

2.   *Spirulina* algal genus in alkaline ponds or sewage water containing high amount of organic matter.

3.   Baker's yeast as protein supplement.

4.   *Torula* yeast (*Candida utilis*) as a protein source for human and animals. Raw material is sulphite waste liquor from pulp and paper manufacture and wood sugar from acid hydrolysis of wood.

Microbial flora of fresh food

The inner tissues of healthy plants and animals are free of microbial activity. However, the surface of raw vegetables and meats are contaminated with a variety of microorganisms.

i.   The carcass of a healthy animal slaughtered for meat and held in refrigerated room is likely to have only nominal surface contamination while the inner tissue are sterile. Each new surface of meat, resulting from a new cut, adds more microorganism to the exposed tissue. Common species of bacteria occurring on fresh meats are *Pseudomonods*, *Staphylococcus*, *Micrococcus*, *Enterococcus and* coliforms. The low temperature at which fresh meats are held, favors the growth of psychrophilic microbes.

ii. The microorganisms found in milk are *Streptococcus lactis, Streptococcus cremoris* and certain other *Lactobacilli*. The principle change is lactose fermentation forming lactic acid that develop sour flavor.

Microbial Spoilage of Food and Food Poisoning

Given a chance to grow, the organism will produce changes in appearance, flavour, odour and other qualities of foods. This degradation process may be described as follows

a. Putrefaction

Protien food + Proteolytic M.O ------→Amino acids + amines

+ ammonia + hydrogen sulfide

b. Fermentation

Carbohydrates rich food + microbes ------→Acids + Alcohals + gases

or methane and carbon dioxide

c. Rancidity

Fattyacids + lipolytic microorganisms ------→Smaller fatty acids +

glycerol + CO_2

Food serves as an excellent media for microbes, because food contains proteins, carbohydrates and fat etc. The action of proteolytic microbes on proteinaceous food under anaerobic conditions results in production of off flavors (due to presence of hydrogen sulfide). It is known as putrefaction. Carbohydrates in food are converted into acids, alcohol and gases (carbon dioxide) due to fermentation. Fats are broken down to smaller fatty acids and glycerol. The common types of spoilage and microbes involved are as follows

Table 14.2: Common types of spoilage and microbes

Food	Types of spoilage	Microbes involved
Bread	Moldy	*Rhizopus, Penicillium, Aspergillus.*
Fruits and vegetables	Ropy sour	*Bacillus subtilis, Lactobacillus.*
Fruits and vegetables	Soft rot, gray mold rot and black mold rot.	*Rhizopus, Erwinia, Aspergillus, Botrytis.*
Meat and Fish	Putrefaction	*Bacillus, Alkaligens, Clostridium, Pseudomonas.*
Concentrated juices	Off flavour	*Lactobacillus*
Ropiness in milk	Off flavour	*Bacillus, Alkaligens, Clostridium, Streptococcus*
Canned foods	Putrefaction	*Clostridium, Bacillus, Micrococcus.*

Table 14.3: Various foods spoiled by microbes

Food	Types of spoilage	Microbes involved
Fresh fruit and vegetables	Soft rot	*Rhizopus, Erwinia,*
	Gray mold rot	*Botrytis*
	Black mold rot	*Aspergillus niger*
Pickles, Sauerkraut	Film Yeasts	*Rhodotorula*
Fresh meat	Putrefaction	*Clostridium, Proteus*
Cured meat	Moldy	*vulgaris, Pseudomonas*
		Aspergillus, Rhizopus
Fish	Discoloration	*Pseudomonas*
Eggs	Green rot	*Pseudomonas*
	Black rots	*Fluorescens*
		Alcaligenes
Concentrated orange juice	Off flavor	*Lactobacillus,*
		Acetobacter
Poultry	Slime, odor	*Alcaligenes*
		Pseudomonas

Preservation of foods

a. Fermented dairy products

b. Other fermented foods

(a) Fermented dairy products

In the dairy industry, fermented milks are produced by inoculating pasteurized milk with a known culture of microorganism, sometimes referred to as **starter culture**. Starter cultures include bacteria like *Streptococcus, Thermophilus and Lactobacillus bulgarics*.

A. Milk ⟶ *Streptococcus lactis or Lactobacillus lactis* ⟶ Curd

B. Milk ⟶ Mechanical method ⟶ Churning cream ⟶ *Streptococcus lactis* ⟶ Butter

Leuconostoc citrovorum add for flavors

C. Milk *Streptococcus lactis* Fermented *Lactobacillus bulgaris* Yoghurt milk

D. Milk *Streptococcus lactis* curd *Lactobacillus lactis*

Cheese *Streptococcus lactis*

The toxin production in foods and food poisoning

The term food poisoning refers to the illness caused by the presence of a bacterial toxin formed in food or illness caused by the entrance of bacteria into the body through ingestion of contaminated food. Various toxins are as follows:

(a) Endotoxin is that toxin which is retained within the cells and liberated on bacterial cell lysis. *e.g., E. coli, Salmonella, Schigella, Bacillus cereus.*

(b) Exotoxin is that which is excreted into the medium or surroundings where bacteria are growing *e.g. Clostridium botulinum.*

(c) Mycotoxins are produced by fungi. Its maximum occurrence is in grain. Various examples of mycotoxins are

- Patulin produced by *Penicilline patulum*

- Penicillic acid produced by *Penicillium puberulum*

- Ergot in Bajra produced by *Claviceps purpurea* causes abortion in animals. It has medicinal value also.

(d) **Aflatoxins** are produced by *Aspergillus flavus.*

Table 14.4: Common diseases and their symptoms caused by food poisoning

Disease	Infection	Causative organism	Symptoms
Salmonellosis	+	*Salmonella*	Diarrhoea, abdominal pain, vomiting, fever, dehydration.
Traveller's sickness	+	*E.coli*	Diarrhoea, fever, dehydration.
Schigellosis	+	*Schigella*	Diarrhoea, headache, vomiting, dehydration.
Gastroenteritis	+	*Bacillus cereus*	Diarrhoea, abdominal pain, vomiting, nausea, dehydration.
Botulism	Toxin production	*Clostridium botulinum*	Diarrhoea, headache, dizziness, difficulty in speaking, progressive weakness, ultimately cardiac and respiratory failure.
Food poisoning	+	*Staphylococcus aureus*	Collapse, vomiting, nausea, dehydration.

(b) Other fermented foods

Important food items produced in whole or in part by microbial fermentations include pickles, sauerkraut, olives and certain types of sausage. Lactic acid bacteria are mainly responsible for the desirable type of fermentation required for the production of each of these products.

(c) Fruits and Vegetables

Fruits and vegetables are normally susceptible to infection by bacteria, fungi and viruses. A main factor contributing to the microbial contamination of fruits and vegetables pertains to their post harvest handling. Mechanical handling is likely to produce breaks in the tissue which facilitates invasion by micro-organisms. The pH range for vegetables is slightly higher (pH 5.0 to 7.0) than fruits (2.3 to 5.0), hence they are more susceptible than fruits to be attacked by bacteria.

(d) Milk

The number of bacteria present at the time of milking has been reported to range between several hundreds to several thousands per milliliter. From the time the milk leaves the udder until it is dispensed into containers, everything with which it comes in contact is a potential source of contamination.

The various practices used for food preservation are

- Aseptic handling
- High temperature
- Low temperature
- Dehydration
- Osmotic pressure
- Chemicals
- Radiation

Aseptic handling

Food items undergo considerable handling prior to being processed by some specific method of preservation such as canning, freezing or dehydration.

High Temperatures

High temperature is one of the safest and most reliable methods of food preservation. Heat is widely used to destroy organism in food products in cans, jars and other types of containers that restrict the entrance of micro-organisms after processing. The various procedures are as follows:

- **Steam under pressure** - Steam kills all vegetative cells and spores.
- **Canning** – The temperatures used for canning foods ranges from 100°C for high acid foods to 121°C for low acid foods.
- **Pasteurization of Milk** – Heating milk or milk product to at least 145° F and holding it continuously at or above this temperature for at least 30 minutes or to at least 161°F for 15 minutes.
- **Sterilization** – Milk sterilization techniques have been developed

which expose milk to ultra high temperatures for very short periods of time for example 300°F (148.9°C).

Low temperatures

Temperature approaching 0°C and lower retards the growth and metabolic activities of micro-organisms. Before freezing, the fresh produce is steamed to inactivate enzymes that would alter the product even at low temperature. Quick freeze method, using temperature of –32°C is generally employed.

Preservation of food

The various practices utilized for food preservation may be summarized as follows:

1. Aseptic handling
2. Heat
   a. Boiling
   b. Steam under pressure
   c. Pasteurization.
3. Low temperatures
   a. Refrigeration
   b. Freezing
4. Dehydration
5. Osmotic and pressure
   a. In concentrated sugar
   b. With brine
6. Chemicals
   a. Organic acids
   b. Substances developing during processing (smoking)
   c. Substances contributed by microbial fermentation (acids)
7. Radiation
   a. Ultra violet
   b. Ionizing radiations.

All methods of food preservation are based upon following principles
 i. Prevention or removal of contamination.
 ii. Inhibition of microbial growth and metabolism (microbistatic action).
iii. Killing of microbes (microbicidal action).

Canning

Steam under pressure is the most effective method, since it kills all vegetative cells and spores.

Preservation by chemicals

Among the most effective chemicals acceptable for food preservation are benzoic, sorbic, acetic, lactic and propionic acids. Sorbic and propionic acids are used to inhibit mold growth in bread. Nitrates and nitrites used in curing meats are inhibitory to some anaerobic bacteria. Food prepared by fermentation processes *e.g.* sauerkraut, pickles and silage for animals are preserved mainly by acetic, lactic and propionic acids produced during microbial fermentation. Smoking generates cresols and other antibacterial compounds which penetrate the meat.

Dehydration

The removal of water by drying in the sun and air or with applied heat causes dehydration. The preservative effect of dehydration is mainly due to microbiostasis. The micro-organisms are not necessarily killed. Growth of all organisms can be prevented by reducing the moisture content of their environment below a critical level.

Osmotic Pressure

Water is withdrawn from micro-organisms by adding large amounts of dissolved substances such as sugar or salt to the medium. The cells are plasmolysed and metabolism is arrested. Jellies and jams are rarely affected by bacterial action because of high sugar content. High osmotic pressure may inhibit microbial growth and thus food products are preserved at room temperature for long time.

Chemicals

Only a few chemicals are largely acceptable for food preservation. Among the most effective are benzoic, ascorbic, acetic, lactic and propinic acids; all of which are organic acids. Nitrates and nitrites are used in curing meat while ascorbic and propinic acids are used to inhibit mold growth in bread.

Radiation

Ultraviolet light of sufficient intensity and time of exposure are microbicidal to exposed micro-organisms. Ultraviolet radiation is limited to control of micro-organisms on surface. Ionizing radiations are lethal to micro-organisms, the ability to penetrate are characteristics that make them attractive for control of micro-organisms in foods. Gamma rays and electron beams are extensively used in the food industry.

Chapter 15

Biotechnology

The term is derived from the fusion of biology and technology. Exploitation of biological agents or their components to produce useful products/services is called biotechnology. The other definitions are:

"The controlled use of biological agents, such as microorganism or cellular components for beneficial use" - US National science foundation

"The integrated use of biochemistry, microbiology and engineering sciences in order to achieve technological application of the capabilities of Micro-organism, cultured tissue and parts thereof" - European federation of Biotech

"The application of biological organism, system or processes" - British biotechnology

Biotechnology is the use of living systems and organisms to develop or make useful products, or "any technological application that uses biological systems, living organisms or derivatives thereof, to make or modify products or processes for specific use". For thousands of years, humankind has used biotechnology in agriculture, food production and medicine. The term itself is largely believed to have been coined in 1919 by Hungarian engineer *Karl Ereky*. In the late 20th and early 21st century, biotechnology has expanded to include new and diverse sciences such as *genomics, recombinant gene technologies, applied immunology*, and development of pharmaceutical therapies and diagnostic tests.

The concept of 'biotech' or 'biotechnology' encompasses a wide range of procedures (and history) for modifying living organisms according to human purposes — going back to domestication of animals, cultivation of plants, and "improvements" of them through breeding programs that employ *artificial selection* and *hybridization*. Modern usage also includes *genetic engineering* as well as *cell* and *tissue culture* technologies. Biotechnology is defined by the American Chemical Society as the application of biological organisms, systems, or processes by various industries to learning about the science of life and the improvement of the value of materials and organisms such as pharmaceuticals, crops, and livestock. In other words, biotechnology can be defined as the mere application of technical advances in life science to develop commercial products. Biotechnology includes pure biological sciences (*genetics, microbiology, animal cell culture, molecular biology, biochemistry, embryology, cell biology*). And in many instances it is also dependent on knowledge and methods from outside the sphere of biology including:

* *Chemical engineering*,
* *Bioprocess engineering*,
* *Bioinformatics*,
* *Biorobotics*.

Brief History

Biotechnology has led to the development of antibiotics. In 1928, *Alexander Fleming* discovered the mold *Penicillium*. His work led to the purification of the antibiotic by Howard Florey, Ernst Boris Chain and Norman Heatley, *penicillin*. In 1940, penicillin became available for medicinal use to treat bacterial infections in humans.

The field of modern biotechnology is generally thought of as having been born in 1971 when Paul Berg's (Stanford) experiments in gene splicing had early success. Herbert W. Boyer (Univ. Calif. at San Francisco) and Stanley N. Cohen (Stanford) significantly advanced the new technology in 1972 by transferring genetic material into a bacterium, such that the imported material would be reproduced. The commercial viability of a biotechnology industry was significantly expanded on June 16, 1980, when the *United States Supreme Court* ruled that a *genetically modified microorganism* could be *patented* in the case of *Diamond v. Chakrabarty*. Indian-born Ananda Chakrabarty had modified a bacterium (*Pseudomonas putida*) capable of breaking down crude oil, which he proposed to use in treating oil spills. (Chakrabarty's work did not involve gene manipulation but rather the transfer of entire organelles between strains of the *Pseudomonas* bacterium.

Rising demand for biofuels is expected to be good news for the biotechnology sector, with the *Department of Energy* estimating *ethanol* usage could reduce U.S. petroleum-derived fuel consumption by up to 30 per cent by 2030. The biotechnology sector has allowed the U.S. farming industry to rapidly increase its supply of corn and soybeans—the main inputs into biofuels—by developing genetically modified seeds which are resistant to pests and drought. By boosting farm productivity, biotechnology plays a crucial role in ensuring that biofuel production targets are met.

Applications

Biotechnology has applications in four major industrial areas, including health care (medical), crop production and agriculture, non food (industrial) uses of crops and other products (e.g. *biodegradable plastics, vegetable oil, biofuels*), and environmental uses.

For example, one application of biotechnology is the direct use of *organisms* for the manufacture of organic products (examples include *beer* and *milk* products). Another example is using naturally present *bacteria* by the mining industry in *bioleaching*. Biotechnology is also used to recycle, treat waste, cleanup sites contaminated by industrial activities (*bioremediation*), and also to produce *biological weapons*.

A series of derived terms have been coined to identify several branches of biotechnology; for example:

- *Bioinformatics* is an interdisciplinary field which addresses biological problems using computational techniques, and makes the rapid organization and analysis of biological data possible. The field may also be referred to as *computational biology*, and can be defined as, "conceptualizing biology in terms of molecules and then applying informatics techniques to understand and organize the information associated with these molecules, on a large scale." Bioinformatics plays a key role in various areas, such as *functional genomics, structural genomics*, and *proteomics*, and forms a key component in the biotechnology and pharmaceutical sector.

- *Blue biotechnology* is a term that has been used to describe the marine and aquatic applications of biotechnology, but its use is relatively rare. It mainly deals with fish production.

- *Green biotechnology* is biotechnology applied to agricultural processes. An example would be the selection and domestication of plants via *micropropagation*. Another example is the designing of *transgenic plants* to grow under specific environments in the presence (or absence) of chemicals. One hope is that green biotechnology might produce more environmentally friendly solutions than traditional *industrial agriculture*. An example of this is the engineering of a plant to express a *pesticide*, thereby ending the need of external application of pesticides. An example of this would be *Bt corn*. Whether or not, green biotechnology products such as this are ultimately more environmentally friendly is a topic of considerable debate. Another example is Bt cotton; in which if gene from *Bacillus thuriengiensis* has been introduced which is resistant to ball worm insect.

- *Red biotechnology* is applied to medical processes. Some examples are the designing of organisms to produce *antibiotics*, and the engineering of genetic cures through *genetic manipulation*.

- *White biotechnology*, also known as industrial biotechnology, is biotechnology applied to *industrial* processes. An example is the designing of an organism to produce a useful chemical. Another example is the using of *enzymes* as industrial *catalysts* to either produce valuable chemicals or destroy hazardous/polluting chemicals. White biotechnology tends to consume less in resources than traditional processes used to produce industrial goods.

2. Medicine

In medicine, modern biotechnology finds promising applications in such areas as:

- *drug production*
- *pharmacogenomics*
- *gene therapy*

1. Better vaccines: Safer vaccines can be designed and produced by organisms transformed by means of genetic engineering. These vaccines will elicit the immune response without the attendant risks of infection. They will be inexpensive, stable, easy to store, and capable of being engineered to carry several strains of pathogen at once.

Modern biotechnology is often associated with the use of genetically altered *microorganisms* such as *E. coli* or *yeast* for the production of substances like synthetic *insulin* or *antibiotics*. It can also refer to *transgenic animals* or *transgenic plants*, such as *Bt corn*. Genetically altered mammalian cells, such as *Chinese Hamster Ovary cells* (CHO), are also used to manufacture certain pharmaceuticals. Another promising new biotechnology application is the development of *plant-made pharmaceuticals*.

Biotechnology is also commonly associated with landmark breakthroughs in new medical therapies to treat *hepatitis B, hepatitis C, cancers, arthritis, haemophilia, bone fractures, multiple sclerosis,* and *cardiovascular* disorders. The biotechnology industry has also been instrumental in developing molecular diagnostic devices that can be used to define the target patient population for a given biopharmaceutical. *Herceptin*, for example, was the first drug approved for use with a matching diagnostic test and is used to treat breast cancer in women whose cancer cells express the protein *HER2*.

Modern biotechnology can be used to manufacture existing medicines relatively easily and cheaply. The first genetically engineered products were medicines designed to treat human diseases. To cite one example, in 1978 *Genentech* developed synthetic humanized *insulin* by joining its gene with a *plasmid* tiplasmod vector inserted into the bacterium *Escherichia coli*. Insulin, widely used for the treatment of diabetes, was previously extracted from the pancreas of *abattoir* animals (cattle and/or pigs). The resulting genetically engineered bacterium enabled the production of vast quantities of synthetic human insulin at relatively low cost.

2. Gene therapy: Gene therapy using an *Adenovirus* vector. A new gene is inserted into an adenovirus vector, which is used to introduce the modified *DNA* into a human cell. If the treatment is successful, the new gene will make a functional *protein*.

Gene therapy may be used for treating, or even curing, genetic and acquired diseases like cancer and AIDS by using normal genes to supplement or replace defective genes or to bolster a normal function such as immunity. It can be used to target *somatic cells* (i.e., those of the body) or *gamete* (i.e., egg and sperm) cells. In somatic gene therapy, the genome of the recipient is changed, but this change is not passed along to the next generation. In contrast, in germline gene therapy, the egg and sperm cells of the parents are changed for the purpose of passing on the changes to their offspring.

There are basically two ways of implementing a gene therapy treatment:

(a) Ex vivo, which means "outside the body" – Cells from the patient's blood or *bone marrow* are removed and grown in the laboratory. They are then exposed to a virus carrying the desired gene. The virus enters the cells, and the desired gene becomes part of the DNA of the cells. The cells are allowed to grow in the laboratory before being returned to the patient by injection into a vein.

(b) **Cloning:** Cloning involves the removal of the nucleus from one cell and its placement in an unfertilized egg cell whose nucleus has either been deactivated or removed.

There are two types of cloning:

1. Reproductive cloning. After a few divisions, the egg cell is placed into a uterus where it is allowed to develop into a fetus that is genetically identical to the donor of the original nucleus.

2. Therapeutic cloning. The egg is placed into a Petri dish where it develops into embryonic stem cells, which have shown potentials for treating several ailments.

3. Agriculture

Using the techniques of modern biotechnology, one or two *genes* may be transferred to a highly developed crop variety to impart a new character that would increase its yield. However, while increases in crop yield are the most obvious applications of modern biotechnology in agriculture, they are also the most difficult ones. Current genetic engineering techniques work best for effects that are controlled by a single gene. Many of the genetic characteristics associated with yield (e.g., enhanced growth) are controlled by a large number of genes, each of which has a minimal effect on the overall yield. There is, therefore, much scientific work to be done in this area. The other applications in the field of agriculture include introduction of genes for salt tolerance disease resistance and watch stress.

Reduced vulnerability of crops to environmental stresses

Crops containing genes that will enable them to withstand biotic and abiotic stresses may be developed. For example, *drought* and excessively salty soil are two important *limiting factors* in crop productivity. Biotechnologists work to find genes that enable some plants to cope with these extreme conditions and eventually to transfer these genes to the more productive crops. One of the latest developments is the identification of a plant gene, At-DBF2, from *Arabidopsis thaliana*. *Arabidopsis thaliana* is a tiny weed often used for plant research because it is very easy to grow. Its genetic code, approximately 115 Mb of the 125 Mb genome, which has been sequenced and interpreted and which can be manipulated in many ways. The At-DBF2 gene shows tolerance to salt, drought, heat and cold in plants. When this gene was inserted into *tomato* and *tobacco* cells the cells withstood these conditions far better than ordinary cells. If these preliminary results prove successful in larger trials, then At-DBF2 genes can help in engineering crops that can better withstand harsh environments. Researchers have also created transgenic rice plants that resist *rice yellow mottle virus* (RYMV). In Africa, this virus destroys a majority of the rice crops and makes the surviving plants more susceptible to fungal infections. While all of these technological advances have the probability for commercial use, they need to be researched more publicly so they can be proven as a stable source of production.

Increased nutritional qualitiesProteins in foods may be modified to increase their nutritional qualities. Proteins in legumes and cereals may be transformed to provide the amino acids needed by human beings for a balanced diet. An example is the work of Professors *Ingo Potrykus* and *Peter Beyer* in creating *Golden rice*. The rice was a result of utilizing genetic modification with genetic material from corn and a soil microorganism. The genetically modified rice produced *beta carotene* which is converted to *vitamin A*. The extra beta carotene content turned the rice a golden colour.

Improved taste, texture or appearance of food

Modern biotechnology can be used to slow down the process of spoilage. Modified fruit can ripen longer on the plant and then be transported to the consumer with less risk of spoilage, and a still-reasonable shelf life. This alters the taste, texture and appearance of the fruit. Reduction in spoilage could expand the market for farmers in developing countries. The first genetically modified food product was a tomato which was transformed to delay its ripening. Biotechnology in cheese production: enzymes produced by micro-organisms provide an alternative to animal rennet – a cheese coagulant – and an alternative supply for cheese makers.

About 85 million tons of wheat flour is used every year to bake bread. By adding an enzyme called *maltogenic amylase* to the flour, bread stays fresher longer. Assuming that 10–15 per cent of bread is thrown away as stale, if it could be kept fresh another 5–7 days then perhaps 2 million tons of flour per year would be saved. Other enzymes can cause bread to expand to make a lighter loaf, or can alter the loaf in a range of ways.

Reduced dependence on fertilizers, pesticides and other agrochemicals

Most of the current commercial applications of modern biotechnology in agriculture are on reducing the dependence of farmers on *agrochemicals*. For example, *Bacillus thuringiensis* (Bt) is a soil bacterium that produces a protein with insecticidal qualities. Traditionally, a fermentation process has been used to produce an insecticidal spray from these bacteria. In this form, the *Bt* toxin occurs as an inactive protoxin, which requires digestion by an insect to be effective. There are several *Bt* toxins and each one is specific to certain target insects. Crop plants have now been engineered to contain and express the genes for *Bt* toxin, which they produce in its active form. When a susceptible insect ingests the transgenic crop cultivar expressing the Bt protein, it stops feeding and soon thereafter dies as a result of the Bt toxin binding to its gut wall. Bt corn is now commercially available in a number of countries to control corn borer (a lepidopteran insect), which is otherwise controlled by spraying (a more difficult process).

Crops have also been genetically engineered to acquire tolerance to broad-spectrum herbicide. The lack of herbicides with broad-spectrum activity and no crop injury was a consistent limitation in crop *weed management*. Multiple applications of numerous herbicides were routinely used to control a wide range of weed species detrimental to agronomic crops. Weed management tended to rely on pre-emergence that is, herbicide were sprayed in response to expected weed infestations rather than in response to actual weeds present. Mechanical cultivation and hand weeding were often necessary to control weeds not controlled by herbicide applications. The introduction of herbicide tolerant crops has the potential of reducing the number of herbicide active ingredients used for weed management, reducing the number of herbicide applications made during a season, and increasing yield due to improved weed management and less crop injury. Transgenic crops that express tolerance to glyphosate, glufosinate and bromoxynil have been developed. These herbicides can now be sprayed on transgenic crops without inflicting damage on the crops while killing nearby weeds.

From 1996 to 2001, herbicide tolerance was the most dominant trait introduced to commercially available transgenic crops, followed by insect resistance. In 2001, herbicide tolerance deployed in soybean, corn and cotton accounted for 77 per cent of the 626,000 square kilometres planted to transgenic crops; Bt crops accounted for 15 per cent; and "stacked genes" for herbicide tolerance and insect resistance used in both cotton and corn accounted for 8 per cent.

Production of novel substances in crop plants

Biotechnology is finding novel uses beyond food. For example, oilseed can be modified to produce fatty acids for detergents, substitute fuels and petrochemicals. Potatoes, tomatoes, rice, tobacco, lettuce, safflowers, and other plants have been genetically engineered to produce insulin and certain vaccines. If future clinical trials prove successful, the advantages of edible vaccines would be enormous, especially for developing countries. The transgenic plants may be grown locally and cheaply. Homegrown vaccines would also avoid logistical and economic problems posed by having to transport traditional preparations over long distances and by having to keep them cold in transit. And since they would be edible, they would not need syringes, which are not only an additional expense in the traditional vaccine preparations but also a source of infections if contaminated.

Controversy

There is another side to the agricultural biotechnology issue. It includes increased herbicide usage and resultant herbicide resistance, "super weeds", residues on and in food crops, genetic contamination of non-GM crops which hurt organic and conventional farmers, etc. Besides this, there is the obvious discomfort and concern of the public concerned with preserving that which has developed on its own in nature for millions of years and seems to have a source of being many consider vital and original and which belongs, and which has superior taste with no equal in quality, character or suitability no matter how large or how small and which no man we know of could have ever invented or devised. In other words, the idea of leaving well enough alone has merit.

Bioengineering

Biotechnological engineering or biological engineering is a branch of engineering that focuses on biotechnologies and biological science. It includes different disciplines such as biochemical engineering, biomedical engineering, bio-process engineering, biosystem engineering and so on. Because of the novelty of the field, bioengineer is still not clearly defined. However, in general it is an integrated approach of fundamental biological sciences and traditional engineering principles.

Biotechnologists are often employed to scale up bio processes from the laboratory scale to the manufacturing scale. Moreover, as with most engineers, they often deal with management, economic and legal issues. Since *patents* and regulation (e.g., U.S. Food and Drug Administration regulation in the U.S.) are very important issues for biotech enterprises, bioengineers are often required to have knowledge related to these issues.

The increasing number of biotech enterprises is likely to create a need for bioengineers in the years to come. Many universities throughout the world are now providing programmes in bioengineering and biotechnology (as independent programs or specialty programs within more established engineering fields).

Biotechnology regulations

The National Institutes of Health (NIH) was the first federal agency to assume regulatory responsibility in the United States. The Recombinant DNA Advisory Committee of the NIH published guidelines for working with recombinant DNA and recombinant organisms in the laboratory. Nowadays, the agencies that are responsible for the biotechnology regulation are: US Department of Agriculture (USDA) that regulates plant pests and medical preparation from living organisms, Environmental Protection Agency (EPA) that regulates pesticides and herbicides, and the Food and Drug Administration (FDA) whose responsibility it is to ensure that the food and drug products are safe and effective.

Operation	Discipline	
Strain choice & improvement	• Microbiology	Zoology
	• Cell biology	Genetics
	• Systematics	
	• Botony	
Mass culture	• Physiology	
Optimization of cell responses	• Process engineering	
Process operation		
Process recovery or	• Chemistry	
Down stream processing	• Biochemistry	
	• Chemical engineering	
	• Biochemical engineering	

In India, Department of Biotechnology under Science and Technology Ministry formulates rules and regulations regarding application of genetically engineered bioproducts for common use.

Relations in biotechnological processes and disciplines which are associated with them. In India, Department of a Biotechnology under:

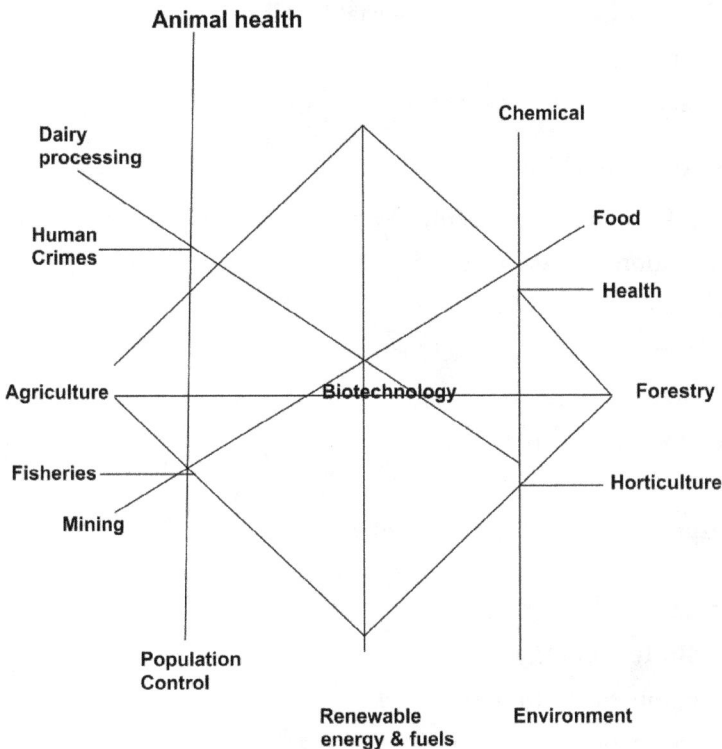

Fig. 15.1. Relations of different platforms on which biotechnology is based

Agricultural Biotechnology

- Clone propagation of trees and plant
- Transfer of novel gene
- Transfer of disease resistance
- Transfer of insect resistance gene *e.g.* cry gene
- Transfer of herbicide resistance gene
  e.g. Glyphosate genes are aroA, bar gene.
- Transfer of bacterial and fungal resistance gene
- Drought resistance.
- Improved of oil quality
- Golden Rice
- Biodegradable plastic (poly-hydroxy butyrate)
- Enzyme – Aceto acetyl CoA reductase and PHB Synthase.

Plant Biotechnology

- Embryo culture to rescue invariable plants
- Germplasm conservation
- Recovery of virus free plant

Environment Biotechnology

- Efficient sewage treatment, deodorization of human excreta
- Degradation of petroleum and management of oil skills
- Detoxification of waste and industrial effluents
- Environmental free pesticides.

Animal Biotechnology

- Test tube babies in human
- *In vitro* fertilization
- Transgenic animal for increased milk, growth rate and production of animal protein.
- Production of

Industrial Biotechnology

- Production of antibiotics
- Production of enzyme
- SCP

- Immobilization of enzymes for their repeated industrial application.
- Protein production

Medical Biotechnology

- Monoclonal antibodies
- DNA Probes
- Recombinant vaccines
- Gene therapy
- Identification of parent/criminal

Enzyme Biotechnology

- Production of enzymes
- Production of abzymes
- Production of ribozymes
- Immobilization of enzyme

Fuel Biotechnology

- Production of Bio-ethanol
- Production of Bio-butanol
- Production of Bio-diesel
- Production of Bio-hydrogen

Terminology

Microorganisms

Microorganisms are the organisms that are to small to be perceived clearly by the unaided human eye i.e., the organisms which have size less than 0.01mm are called microorganisms.

Simple microscope

A microscope having a single lens of a very short focal length.

Compound microscope

The microscope having a double lens system consisting of an ocular and objective lenses.

Tyndallization

A method of sterilization by discontinuous heating.

Fermentation

A process that results in the formation of alcohols or organic acids as a result of breakdown of carbohydrates which are predominant organic compounds in plant tissues.

Putrefaction

A process of decomposition that results in the formation of ill smelling products as a consequence of breakdown of proteins, the principal organic constituents in animals.

Anaerobes

The organisms that can live only in the absence of free oxygen.

Pure culture

A pure culture is one that contains only a single kind of microorganism.

Etiology
Causation of disease.

Surgrical sepsis
The infections that follows surgrical interventions.

Pasteurization
The process of killing of vegetative forms of harmful microorganisms of liquid food or beverages to enhance the keeping quality by heating the food at 72°C for 30 seconds or 72°C for 15 minutes.

Antibiotics
Secondary metabolites produced by microorganisms especially fungi and actinomycetes and kill Gram positive and Gram negative bacteria and some fungal pathogens at low concentration.

Enrichment culture technique
Principal of application on a micro scale of natural selection; use of selective culture media and incubation conditions to isolate microorganisms directly from natural samples.

Sterilization
The process of destroying all forms of microbial life is sterilization. The term sterile, sterilize and sterilization refers to the complete absence or destruction of all microorganisms.

Disinfectant
An agent, usually a chemical that kills the growing forms but not necessarily the resistant spore forms of disease producing microorganisms. Disinfection is the process of destroying infectious agents.

Antiseptic
A substance that opposes sepsis; prevents the growth or action of microorganisms either by destroying microbes or by inhibiting their growth and activity. It is associated with substances applied to animate objects.

Sanitizer
An agent that reduces the microbial population to safe levels. Usually, it is a chemical agent that kills 99.9 per cent of the growing bacteria.

Germicide
An agent that kills the growing forms but not necessarily the resistant spore forms of germs.

Bactericide and bacteriostatic

An agent that kills bacteria is called bactericidal. An agent which inhibits the growth of bacteria is called bacteriostatic. Similar terms are fungicide, virucide and sporicide.

Antimicrobial agents

One that interferes with the growth and activity of microbes is called antimicrobial agent. Some antimicrobial agents are used to treat infections and they are called therapeutic agents.

Enrichment

If the mixed microbial population is introduced into a liquid selective medium, the bacterial culture selective to those conditions will grow. For example isolation of *E. coli* is done by inclusions of the bile salts and growth at 45°C.

Prototrophs

The bacteria which can drive all carbon requirements from the principal carbon source are called prototrophs.

Replication

A process involving in the duplication of the nucleic acid to give identical copies of it.

Transcription

Generate a single strand RNA copying one of the two strands of DNA with the help of RNA polymerase.

Translation

Converts the nucleotide sequence of RNA into the sequence of amino acids composing a protein.

Reverse transcription

It is found in retroviruses. It includes synthesis of DNA using RNA as template with the help of reverse transcriptase.

Polyribosomes or polysomes

An mRNA associated with several ribosomes.

Auxotrophs

In addition to the principal carbon source, one or more organic nutrients are also required for growth of bacteria.

Microaerophilic

The microorganisms which can grow only below of 0.2 atmospheric pressure.

Synthetic medium

A medium that is composed of ingredients of known chemical composition is synthetic medium.

Complex medium

One that contains ingredients of unknown chemical composition is complex medium.

Selective media

Direct isolation- the apacial dispersion of the microbial population on a solid medium considerably reduces the competition for nutrients.

Culture Media

A medium composed entirely of chemically defined nutrients is termed as a synthetic medium. The medium that contains ingredients of unknown chemical composition is termed as complex medium. The separation of a particular microorganism from the mixed populations that exist in nature is called isolation. The growth of microbial populations in artificial environments (culture media) under laboratory conditions is known as cultivation.

References and Notes

1. *Text of the CBD*. Cbd.int. Retrieved on 2013-03-20.

2. *"Incorporating Biotechnology into the Classroom – What is Biotechnology?", from the curricula of the 'Incorporating Biotechnology into the High School Classroom through Arizona State University's BioREACH PROGRAM', accessed on October 16, 2012).* Public.asu.edu. Retrieved on 2013-03-20.

3. *"Incorporating Biotechnology into the Classroom – What is Biotechnology?", from Incorporating Biotechnology into the High School Classroom through Arizona State University's BioREACH PROGRAM, Arizona State University, Microbiology Department, retrieved October 16, 2012.* Public.asu.edu. Retrieved on 2013-03-20.

4. *Biotechnology*. Portal.acs.org. Retrieved on 2013-03-20.

5. *What is biotechnology?*. Europabio. Retrieved on 2013-03-20.

6. *KEY BIOTECHNOLOGY INDICATORS (December 2011)*. oecd.org

7. *Biotechnology policies – Organisation for Economic Co-operation and Development*. Oecd.org. Retrieved on 2013-03-20.

8. *What is Bioengineering?*. Bionewsonline.com. Retrieved on 2013-03-20.

9. See Arnold, John P. (2005) [1911]. Origin and History of Beer and Brewing: From Prehistoric Times to the Beginning of Brewing Science and Technology. Cleveland, Ohio: BeerBooks. p. 34. *ISBN 978-0-9662084-1-2*. OCLC 71834130.

10. [abcd] Thieman, W.J.; Palladino, M.A. (2008). *Introduction to Biotechnology*. Pearson/ Benjamin Cummings. *ISBN 0-321-49145-9*.

11. Springham, D.; Springham, G.; Moses, V.; Cape, R.E. (24 August 1999). *Biotechnology: The Science and the Business*. CRC Press. p. 1. *ISBN 978-90-5702-407-8*.

12. *"Diamond v. Chakrabarty, 447 U.S. 303 (1980). No. 79-139." United States Supreme Court*. June 16, 1980. Retrieved on May 4, 2007.

13. *VoIP Providers And Corn Farmers Can Expect To Have Bumper Years In 2008 And Beyond, According To The Latest Research Released By Business Information Analysts At IBISWorld*. Los Angeles (March 19, 2008).

14. *The Recession List — Top 10 Industries to Fly and Fl... (ith anincreasing share accounted for by ...)*, bio-medicine.org

15. Gerstein, M. *"Bioinformatics Introduction." Yale University.* Retrieved on May 8, 2007.

16. [a] [b] U.S. Department of Energy Human Genome Program, supra note 6.

17. Bains, W. (1987). *Genetic Engineering For Almost Everybody: What Does It Do? What Will It Do?.* Penguin. p. 99. *ISBN 0-14-013501-4.*

18. *IDF 2003; "Diabetes Atlas,: 2nd ed."; International Diabetes Federation, Brussels,* eatlas.idf.org

19. *IDF March 2005; "Position Statement." International Diabetes Federation, Brussels.* idf.org

20. [a] [b] U.S. Department of State International Information Programs, "Frequently Asked Questions About Biotechnology", USIS Online; available from *USinfo.state.gov,* accessed 13 September 2007. Cf. Feldbaum, C. (February 2002). "Some History Should Be Repeated". *Science* 295 (5557): 975. *doi:10.1126/science.1069614. PMID 11834802.*

21. Cameron D. (23 May 2002). "Stop the Cloning". *Technology Review.* Also available from *Techreview.com,* [hereafter "Cameron"]

22. Nussbaum, M.C.; Sunstein, C.R. (1998). *Clones And Clones: Facts And Fantasies About Human Cloning.* New York: W.W. Norton. p. 11. Asian Development Bank, Agricultural Biotechnology, Poverty Reduction and Food Security (Manila: Asian Development Bank, 2001). Also available from *ADB.org*

23. *a b* D. Bruce and A. Bruce, *Engineering Genesis: The Ethics of Genetic Engineering,* London: Earthscan Publications, 1999 *ISBN 1-85383-570-6*

24. *Arabidopsis.* Nih.gov. Retrieved on 2013-03-20.

25. Sara Abdulla (27 May 1999). "Drought stress". *Nature News. doi:10.1038/news990527-9.*

26. National Academy of Sciences (2001). *Transgenic Plants and World Agriculture.* Washington: National Academy Press.

27. *About Golden Rice.* Irri.org. Retrieved on 2013-03-20.

28. For an account of the research and development of Flavr Savr tomato, see Martineau, B. (2001). *First Fruit: The Creation of the Flavr Savr Tomato and the Birth of Biotech Food.* New York: McGraw-Hill.

29. A.F. Krattiger, An Overview of ISAAA from 1992 to 2000, ISAAA Brief No. 19-2000, 9.

30. *EuropaBio — An animal-friendly alternative for cheeze makers,* Europabio.org

31. *EuropaBio — Biologically better bread,* Europabio.org

32. L. P. Gianessi, C. S. Silvers, S. Sankula and J. E. Carpenter. *Plant Biotechnology: Current and Potential Impact for Improving Pest management in US Agriculture, An Analysis of 40 Case Studies* (Washington, D.C.: National Center for Food and Agricultural Policy, 2002), 5–6.

33. C. James, "Global Review of Commercialized Transgenic Crops: 2002", ISAAA Brief No. 27-2002, at 11–12. Also available from *ISAAA.org*

34. Pascual DW (2007). *"Vaccines are for dinner"*. *Proc Natl Acad Sci USA* 104 (26): 10757–10758. *doi:10.1073/pnas.0704516104*. *PMC 1904143*. *PMID 17581867*.

35. *SemBioSys.ca*. SemBioSys.ca. Retrieved on 2011-09-05.

36. Van Eenennaam, AL (2006). *"What is the Future of Animal Biotechnology?"*. *California Agriculture* 60 (3): 132–139. *doi:10.3733/ca.v060n03p132*.

37. Staff (26 December 2012) *Draft Environmental Assessment and Preliminary Finding of No Significant Impact Concerning a Genetically Engineered Atlantic Salmon; Availability* Federal Register / Vol. 77, No. 247 / Wednesday, December 26, 2012 / Notices, Retrieved 2 January 2013.

38. Dove, AW (2005). "Clone on the Range:What Animal Biotech is Bringing to the Table". *Nature Biotechnology* 23 (3): 283–285. *doi:10.1038/nbt0305-283*. *PMID 15765075*.

39. *Monsanto and the Roundup Ready Controversy*, — SourceWatch.org

40. *Monsanto*, — SourceWatch.org

41. Diaz E (editor). (2008). *Microbial Biodegradation: Genomics and Molecular Biology* (1st ed.). Caister Academic Press. *ISBN 1-904455-17-4*.

42. Martins VAP (2008). *"Genomic Insights into Oil Biodegradation in Marine Systems"*. *Microbial Biodegradation: Genomics and Molecular Biology*. Caister Academic Press. *ISBN 978-1-904455-17-2*.

43. *Nigms.Nih.Gov*. Nigms.Nih.Gov. Retrieved on 2011-09-05.

Retrieved from *"http://en.wikipedia.org/w index.php? title= Biotechnology &oldid =562357015"Categories:*

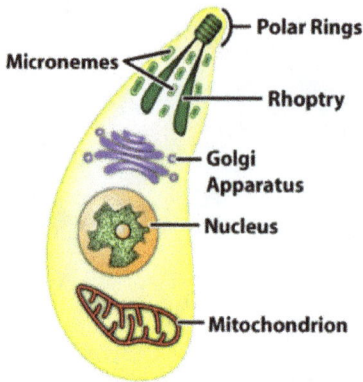

Fig. 2.1. Structure of Protzoan (p. 11)

Fig. 2.2. A view of algae (p. 11)

Fig. 2.3. Fruiting bodies of Fungi (p. 12)

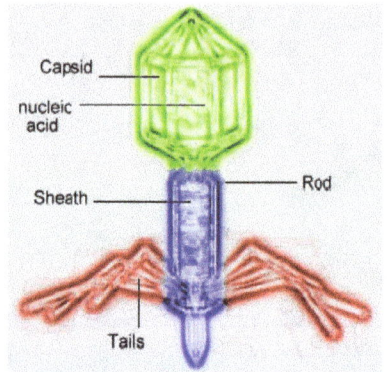

Fig. 2.5. Structure of a Virion (p. 13)

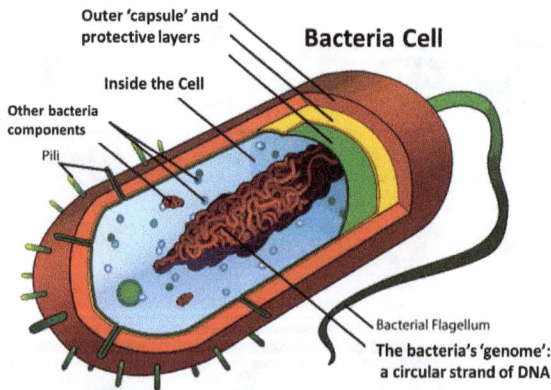

Fig. 2.4. Structure of bacterial cell (p. 13)

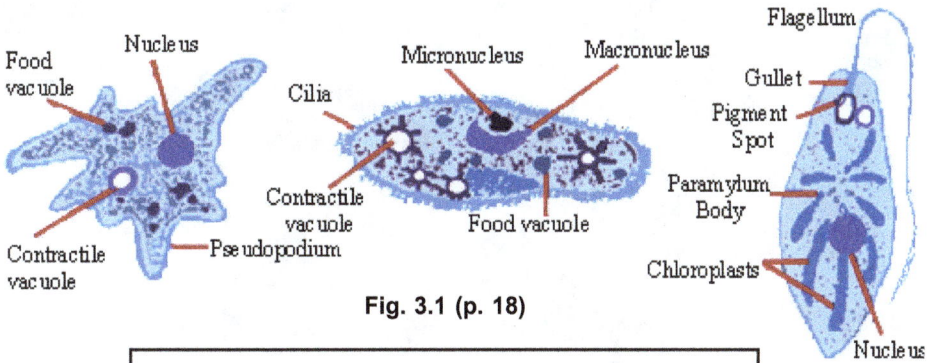

Fig. 3.1 (p. 18)

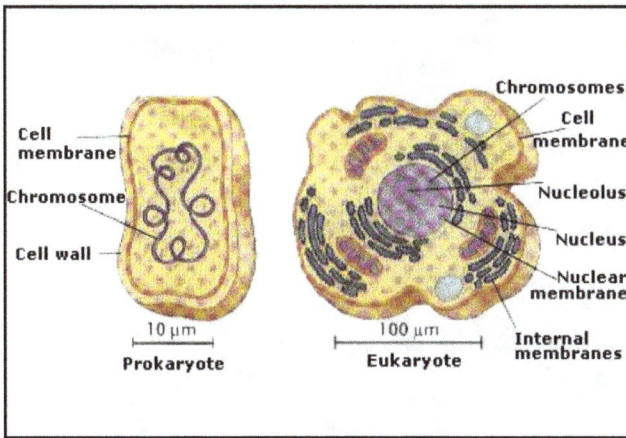

Fig. 3.2. Prokaryotic and Eukaryotic cells (p. 18)

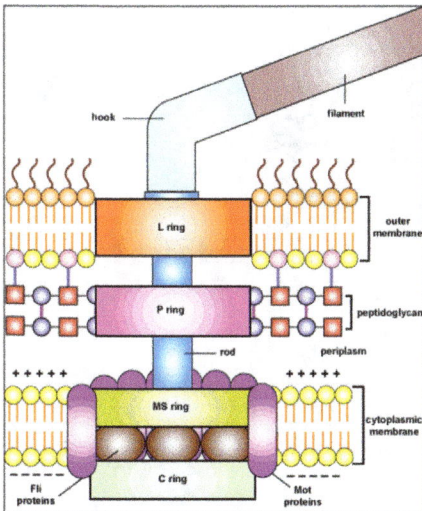

Fig. 3.6. Flagellum of
Gram – ve Bacteria (p. 20)

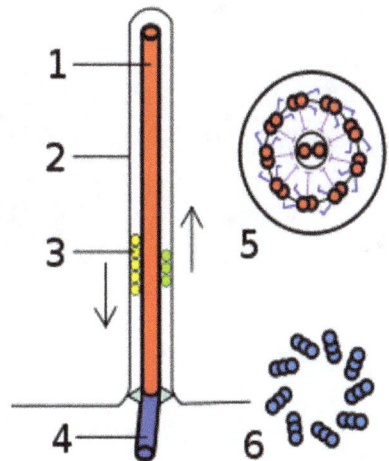

Fig. 3.8. Structure of
Eukaryotic Flagellum (p. 22)

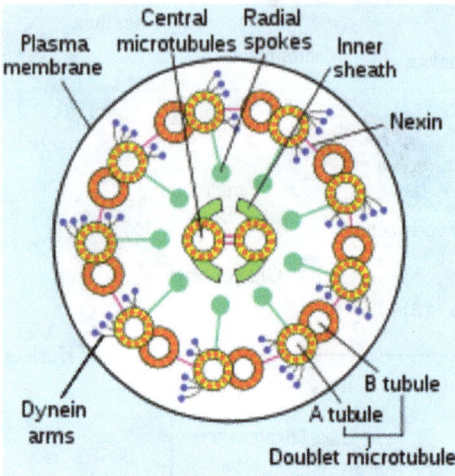

Fig. 3.9. Cross section of an axoneme (p. 22)

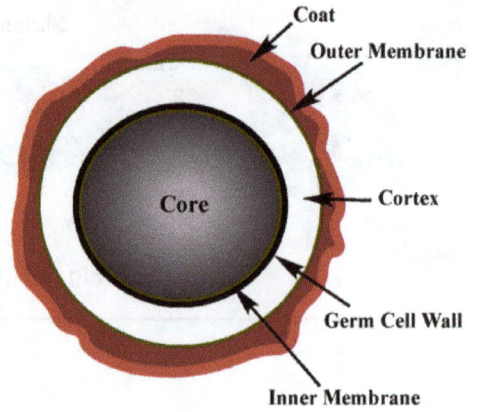

Fig. 3.11. Structure of a bacterial spore (p. 27)

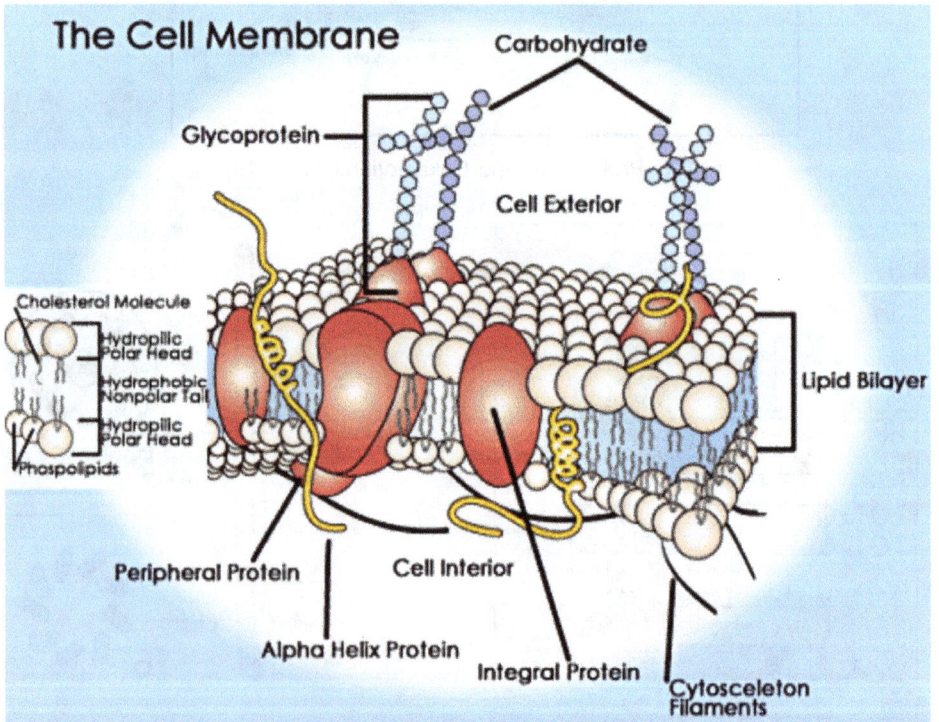

Fig. 3.13: The Cell Membrane (p. 29)

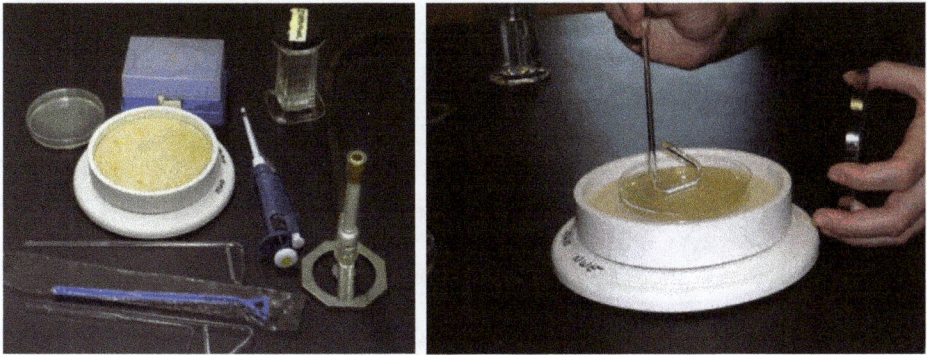

Fig. 4.4. Spread Plate Method (p. 35)

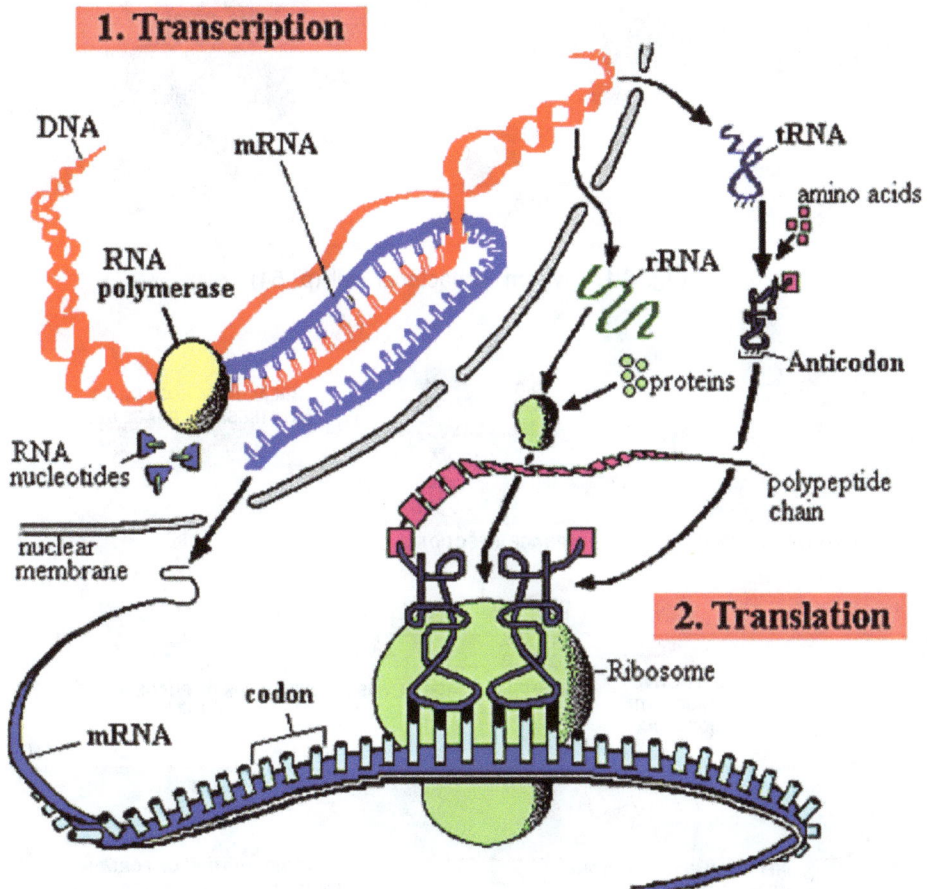

Fig. 5.5. Protein synthesis (p. 50)

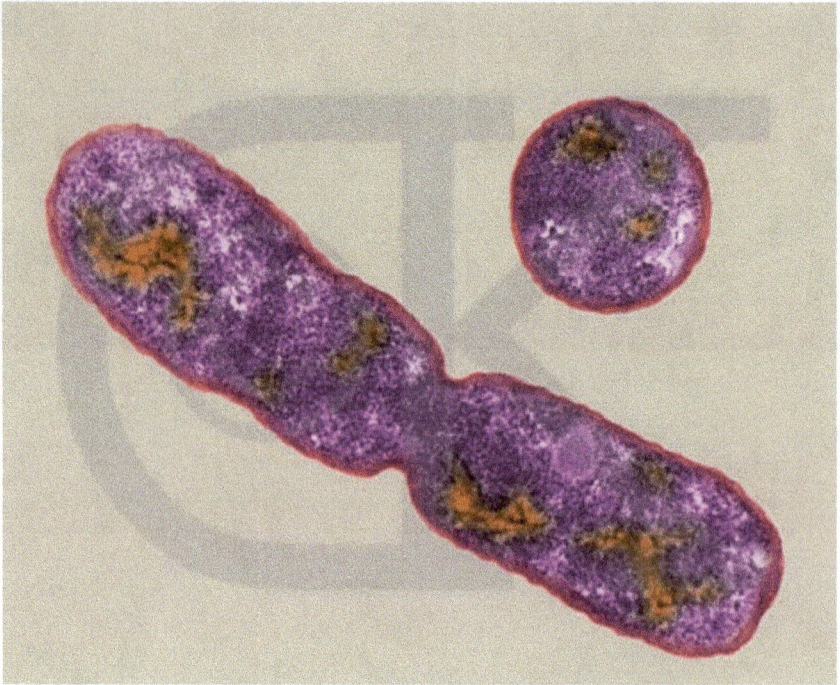

Colour Plates

Fig. 6.1. Division in bacterial cell (p. 54)

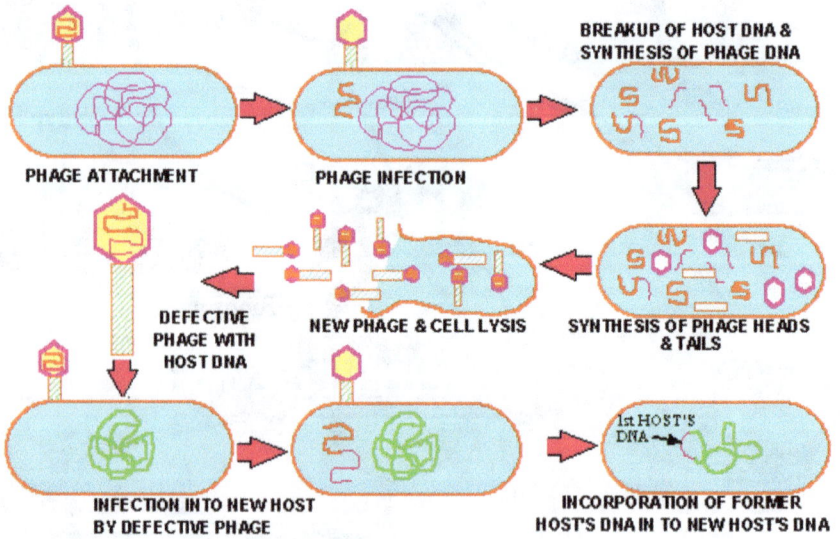

Fig. 6.5. λ phage in E. coli. (p. 59)

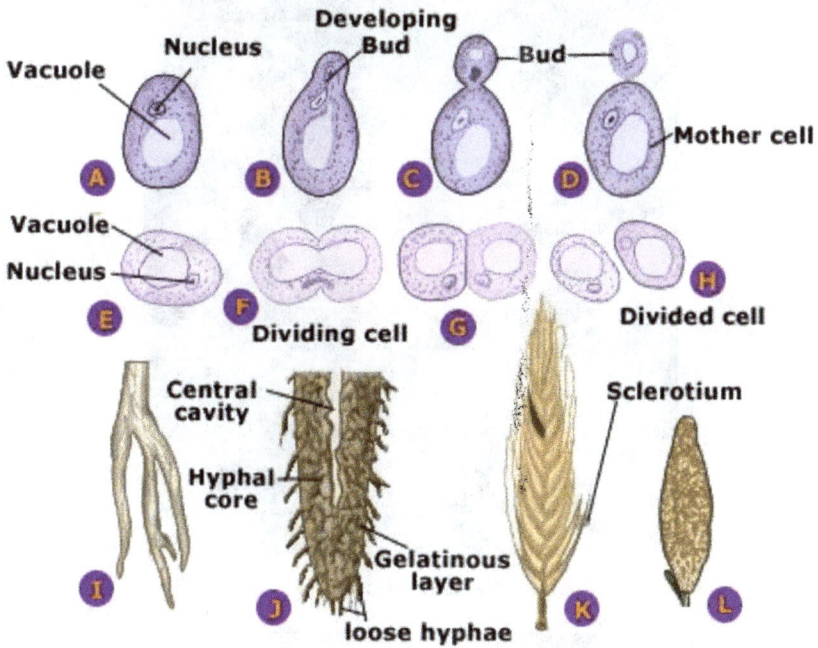

Fig. 6.6. Modes of vegetative reproduction in fungi.
A-D. Budding; E-H. Fission; I-J. Rhizomorph; K-L Scterotia. (p. 59)

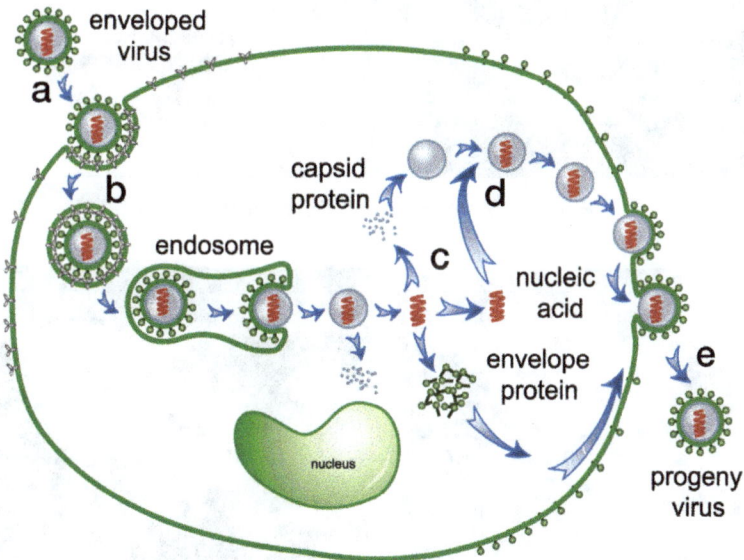

Fig. 7.5. Life cycle of Plant and Animal Viruses (p. 68)

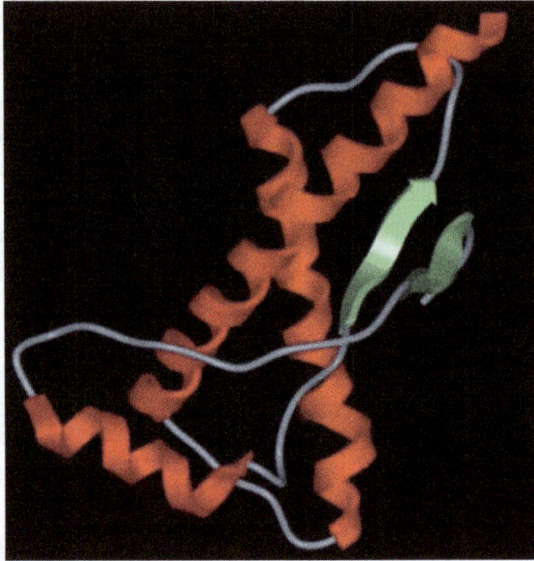

Fig. 7.7. A molecular model of a prion (p. 69)

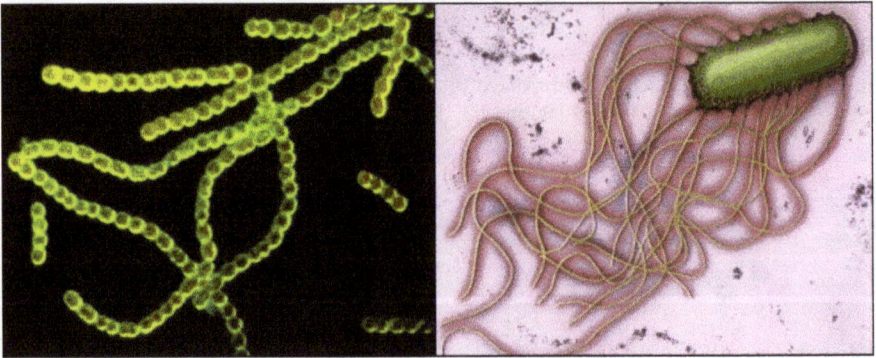

Fig. 8.1. Photosynthetic Bacteria (p. 72)

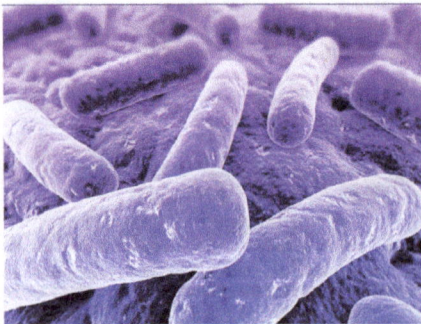

Fig. 8.2. Purple Bacteria (p. 73)

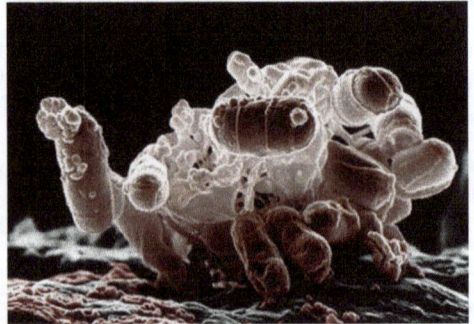

Fig. 8.3. Nitrosomonas (p. 73)

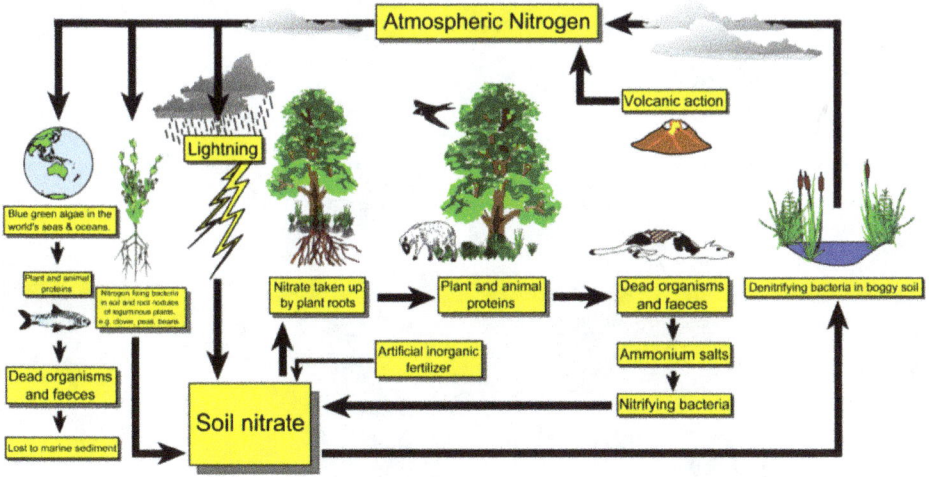

Fig. 9.4. Graphic Sketch of Nitrogen Cycle (p. 89)

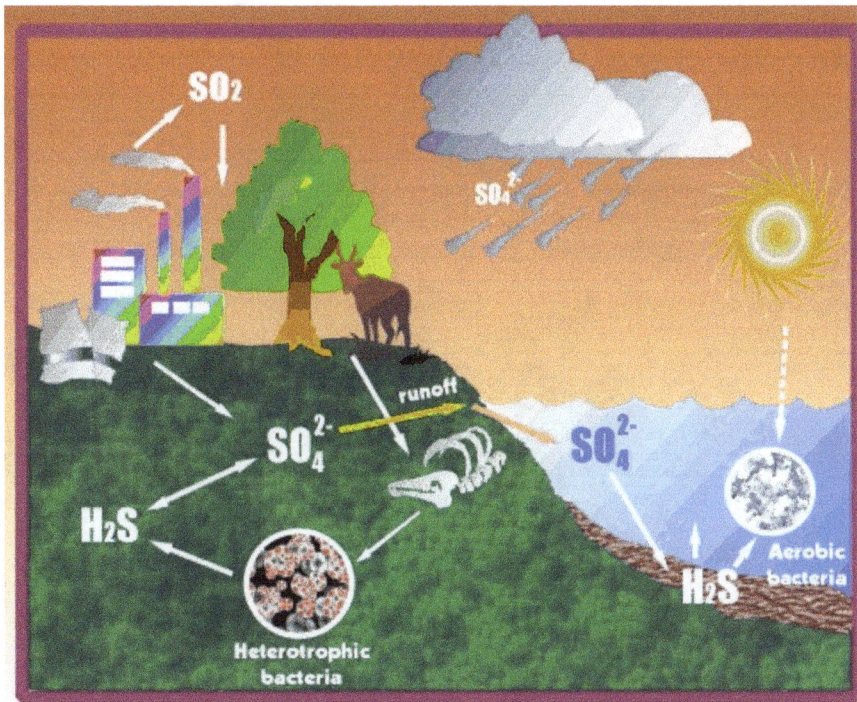

Fig. 9.7. Sulphur Cycle   (p. 96)